Reconstructing School Mathematics

Studies in the Postmodern Theory of Education

Joe L. Kincheloe and Shirley R. Steinberg
General Editors

Vol. 160

PETER LANG
New York • Washington, D.C./Baltimore • Bern
Frankfurt am Main • Berlin • Brussels • Vienna • Oxford

Stephen I. Brown

Reconstructing School Mathematics

Problems with Problems and the Real World

PETER LANG
New York • Washington, D.C./Baltimore • Bern
Frankfurt am Main • Berlin • Brussels • Vienna • Oxford

Library of Congress Cataloging-in-Publication Data
Brown, Stephen I.
Reconstructing school mathematics: problems
with problems and the real world / Stephen I. Brown.
p. cm. — (Counterpoints; vol. 160)
Includes bibliographical references and index.
1. Mathematics—Study and teaching. I. Title.
II. Counterpoints (New York, N.Y.); vol. 160.
QA11 .B815   510'.71—dc21   00-033101
ISBN 0-8204-5103-7
ISSN 1058-1634

Die Deutsche Bibliothek-CIP-Einheitsaufnahme
Brown, Stephen I.:
Reconstructing school mathematics: problems
with problems and the real world / Stephen I. Brown.
–New York; Washington, D.C./Baltimore; Bern;
Frankfurt am Main; Berlin; Brussels; Vienna; Oxford: Lang.
(Counterpoints; Vol. 160)
ISBN 0-8204-5103-7

Cover design by Dutton & Sherman Design

© 2001 Peter Lang Publishing, Inc., New York

All rights reserved.
Reprint or reproduction, even partially, in all forms such as microfilm,
xerography, microfiche, microcard, and offset strictly prohibited.

Dedication

This book is dedicated to

- The people of the village of Le Chambon, who, during the Nazi occupation of Southern France, performed heroic deeds for individuals who were under siege. What is extraordinary is not only that they barely knew the people upon whom they bestowed their caring acts, and not only that they placed their *own* lives in jeopardy by such behavior, but that they sought no rationale for so doing beyond the realization that human beings were being treated in unthinkable ways and that they were in a position to help. Their behavior serves as a reminder that scholarship in the domain of caring is not a *necessary* precondition for the performance of caring deeds.

- Scholars (and "near relatives") who analyze the concept of caring (and "near relatives"), but continue to operate in uncaring ways. I am grateful for their life sustaining reminder that no amount of prior scholarship can merely be "applied" to uncharted territory in critically human domains. Furthermore, they serve as a reminder that mediocre, sloppy, and uncaring analysis is purchased with ease whenever a vigilant grip on context and nuance is relinquished.

- Eileen Brown who demonstrates through every fiber of her being that extraordinary analytical competence and phenomenally caring behavior need not be strange bedfellows.

Contents

List of Illustrations	ix
List of Tables	xi
Preface	xiii
Acknowledgments	xxi

PART I	**Backdrop and Perspective**	1
Chapter 1	Reform in School Mathematics: A Backdrop and Critical Overview	3
	A Problem Solving Perspective	5
	Applications	8
	A Critical Perspective	10
	A Humanistic Perspective	18
	Organization of the Book	30
PART II	**Problem Solving, Problem Posing, and Problems**	33
Chapter 2	Problem Solving and Its Emerging Companion: Philosophical and Educational Roots of Problem Posing	35
	Centrality of Problem Solving in Mathematics	36
	Pedagogical Models Predating the Reform Movement	40
	Problem Posing: An Implicit and Emerging Voice	43
	Problem Posing: Initial Explorations of Its Significance	45
	What-If-Not and Problem Posing: Philosophical Issues	50
	What-If-Not: Educational Matters	55
	What-If-Not: A Strategy	64
Chapter 3	The Concept of Problem and Its Educational Fallacy	69
	The Concept of Problem: Demand and Constraints	70
	The Fallacy	76

Chapter 4 Problem Posing/Solving in a Humanistic Light: Softening the Fallacy	81
Scope and Intentionality	83
Problem/Situation Reverberation	88
Problem/Situation Reverberation: A Return to Progress in Mathematics	93
Revisiting Constraints as a Component of Problems	95
Tightening the Bond of Problem and Solution: An Ironic Twist	97
Revisiting Problem and Solution as *Given*	103
The Concept of Same and Different	113
PART III Reconstructing Real-World Connections	**137**
Chapter 5 An Enhanced View of Connecting Mathematics With the Real World	139
The Established Concept of Connection	139
Alternative Conceptions of Connections	142
Yet Another Scheme	144
Taking Stock and Looking Forward	181
Chapter 6 Problems and the Real World: Conclusion and Expansion	183
Some Tensions Between Abstract and Concrete	184
Back to What Is Real	189
Reality and Mathematics/Education	191
A Retrospective on Problems and Connections	192
Three Resituated Perspectives	196
"Novel" Endings	202
Appendixes	**223**
Appendix A: Talmudic Style Course Disclosure	225
Appendix B: Excerpts From a Mathematical Talmud	227
Notes	235
References	247
Index	265

Illustrations

Figure 2.1	Depicting Gauss's alleged insight for adding numbers from 1 to 100	37
Figure 2.2	Postulate 5 of Euclid's Elements	47
Figure 3.1	Problems traditionally seen as objects to be solved	70
Figure 4.1	Readying the traditional view of problem solving for criticism	82
Figure 4.2	Context as a dimension of problem as given	88
Figure 4.3	The dynamics of problem and situation	92
Figure 4.4	Numerous paths among situation, problem, solution, and context	93
Figure 4.5	The place of context given both problem and solution	98
Figure 4.6	Joining problem and solution to create situations and problems	104
Figure 4.7	Depicting the derivative as the slope of a tangent line to a curve	106
Figure 4.8	Depicting the definite integral as the area under a curve	107
Figure 4.9	Numbers between 0 and 2 as points on a line	108
Figure 4.10	Depicting lines of vision from the origin in an apple orchard	109
Figure 4.11	A clock with numbers 0, 1, and 2	117
Figure 4.12	Wrapping all natural numbers around the clock of Figure 4.11	118
Figure 4.13	An equilateral triangle with a pinpoint at its center	121
Figure 4.14	Spinning the triangle of Figure 4.13 –120 and 240 degrees clockwise	122

Figure 4.15	Depicting a scheme for simplifying addition of numbers ending in zero	124
Figure 4.16	Depicting the travel of Mordecai and Pablo in problem 2	130
Figure 4.17	Depicting additional information for problem 2	131
Figure 5.1	Standard scheme for a mathematical model of a real world problem	140
Figure 5.2	Making salient the discarding of "irrelevant" information in Figure 5.1	143
Figure 5.3	Seeking relevance of given non-mathematical information in relating mathematics to the real world	143
Figure 5.4	Searching for additional information of a non-mathematical nature in relating mathematics to the real world	144
Figure 5.5	Humanistic connections that relate mathematics to the non-mathematical world	145
Figure 5.6	Geometrical depiction of "3 · 5 almost equals 16" from Table 5.2	151
Figure 5.7	The Koch curve created by continuing to replicate smaller equilateral triangles on the sides of an equilateral triangle	175
Figure 6.1	Graphical representation of the extension in Table 6.1	187
Figure 6.2	Schematic representation of a Talmudic page	218

Tables

Table 1.1	Dichotomous depictions of the nature of mathematics and of learning	30
Table 2.1	The divisors for each natural number between 1 and 25	60
Table 2.2	The divisors for each number of E between 1 and 10	61
Table 4.1	The number of primes in several different ranges in N	99
Table 4.2	Addition table for the three sets of Figure 4.12	120
Table 4.3	Summary of the three spins of the equilateral triangle of Figure 13	122
Table 5.1	Doodling with some simple multiplication facts	148
Table 5.2	"Striving to be squares" as a way of seeing Table 5.1	149
Table 5.3	Modifying the doodling of Table 5.1	150
Table 5.4	"Striving to be squares" as a way of seeing Table 5.3	150
Table 5.5	Depicting the "short-fall" of the "striving" of Table 5.4	150
Table 5.6	Revisiting the "striving" of Table 5.4 in search of a constant "short-fall"	151
Table 5.7	Expressing even numbers in N as the sum of two odd numbers	180
Table 6.1	Extrapolating from the product of two positive numbers to find the product of a positive and a negative	185

Preface

This project has been gestating for several years—or has it been several decades? Actually the book itself began quite modestly and serendipitously. I was on leave from the University of Buffalo for half a year in the spring of 1994 as a visiting scholar at Harvard's Philosophy of Education Research Center (PERC)—a small group under the direction of Israel Scheffler and Vernon Howard. Participants were selected annually with an eye towards diversity of interests that related loosely to the study of education. Fields such as architecture, literature, philosophy, science, and mathematics were represented. All members were expected at some point during their residency to make an hour's presentation to the other visiting scholars about a significant issue in their field. I had defined my project during my stay at PERC as that of writing a mathematical novel, but progress was slow, and I was a bit reticent to share my thinking at the time.[1] I therefore scampered about to find an appropriate substitute topic and finally decided to react to a portion of what was advocated in a collection of national documents in mathematics education known as, or associated with, *The Standards.* These documents reflected concern of the profession to make mathematics more accessible to a larger portion of the population and to rectify the conception of mathematics and learning that had been depicted in earlier reforms—from "back to basics" to the "new math." The "new math" had been referred to as a revolution, and it saw proof and axiomatics as the hallmark of post-elementary school mathematics.

I was particularly intrigued with two categories that were central in those documents: a revival of interest in problem solving (especially associated with heuristics) and a concern with "real-world" applications. Focusing upon those categories, I chose to offer what I thought was a modest one-hour presentation to my colleagues. Though sympathetic with the program outlined in *The Standards*, I tried to imagine why problem solving and applications had taken the direction they had, and I wondered what educational alternatives were feasible and desirable. Two major themes influenced the way I thought about both problem solving and "real-world" applications as I prepared the talk: (1) the role of problem *posing*, and (2) a conception of mathematics as a humanistic enterprise.

Though I first developed my thinking about problem posing in the late 1960s—at about the same time that Paulo Freire chose it as a central concept in his educational reform for the oppressed—my agenda was more set on ways of motivating students to do mathematical inquiry and less politically rooted than Freire's program. I began to develop and write about the significance of problem posing and its educational implications, especially for teacher education, in collaboration with my colleague, Marion Walter.

Over the years, I have come to see problem posing as part of a broader educational umbrella—as an invitation to compare the evolution of a culture with the evolution of one's own thinking and feelings. I have come to appreciate its positive value in contributing to even prolonged states of confusion as a significant element in becoming and being educated.

Though *The Standards* make reference to and honor problem posing, they tend to focus less on excavation and expansion of the concept and more on strategies for its implementation. Although *The Standards* are more sophisticated than my earlier association of problem posing primarily as motivation for mathematical inquiry, there has not been an adequate appreciation for its robust educational potential.[2]

Part of what has influenced me to broaden the context has been an evolving interest in the second theme mentioned above: humanistic mathematics (education)—an antidote to mathematics as isolated from other ways of knowing and experiencing the world. I first wrote about the concept in 1973 and have more recently had the opportunity to expand my thinking in this domain partly as a result of my connection with Alvin White—a mathematician from Harvey Mudd College. He organized a conference in 1986 (and subsequently a journal) on humanistic mathematics education. By putting like-minded people in touch with each other from across the mathematical and educational spectrum, he launched a movement with an international audience.

I will speak about these ideas and how they come together in the pages that follow. The short of it, however, is that what I thought would be a one-hour presentation to my colleagues at PERC became a considerably more ambitious project. Two factors were responsible for its growth.

First, due to my postmodern conception of time as well as my disinclination to rehearse talks, I had no way of judging that I would "cover" approximately five percent of what I had planned. I would have been satisfied to think of the other ninety-five percent as both pleasant background music and the frustration one must endure for not taking time as a measurable commodity.

Second my wife, Eileen, a psychologist who has been more aware than I for over forty years of the consequences of my frustrations, was in the audience. As she exemplifies so well herself with her culinary acumen, she persuaded me "to make use of the leftovers." Hence this book, and several years worth of exhilaration/frustration of a different sort.

Author Tensions/Audience Intentions

There were many interesting tensions that I had to acknowledge and work through. The most challenging one was that of defining the style of exposition and the audience as well. Part of the difficulty is that my own orientation is both (1) anecdotal (as this preface exemplifies) and (2) philosophical. While an increasing number of educators are narration friendly, and while there is greater appreciation for philosophical exposition as a way of exploring important educational issues, finding a balance between the two styles of inquiry presents conceptual as well as practical problems.

Furthermore, though I wish to appeal to those who have a commitment to philosophy of mathematics (education), I also hope to capture the imagination of mathematics teachers/educators who agonize over questions of curriculum and teacher education and who engage in debates in the field, but who do not relish philosophical discourse.

Conversely, there is much value in inviting those who have a philosophical orientation outside of mathematics and education to hear something about the controversial issues in the field. Concepts such as problem posing and solving, the meaning of "applying something to the real world," and educational issues related to "understanding" are in need of further criticism and expansion—with the potential for considerable benefit to the mathematics education community.

In addition to the audience of mathematics educators and philosophers, there are educators and interested laymen who might wish to understand and benefit from an awareness of important epistemological type discourse in the realm of mathematics. This is especially so because mathematics has been considered, especially in popular culture, to be the bastion of certainty. To appreciate the existence of a softer perspective in the field of both mathematics and mathematics education offers hope (or for some, dismay) that matters of schooling for educational purposes could be otherwise.

Some comments with regard to grade level and ways of experiencing this book should be helpful. I am not producing a blueprint, but rather an extension/reconception of matters related to teaching/learning mathematics that is not grade specific. Mathematical examples and personal anecdotes, however, are an important part of this document, and I will be selecting them from a wide range of levels of mathematical sophistication—from kindergarten through graduate level. Though I will occasionally make use of specialized mathematical language and symbolism, I will do so sparingly. It will be helpful on a number of occasions, however, for the reader to participate in the actual doing of mathematics in order to appreciate more fully my talking about a reconstructed view of mathematical experiences. Some readers may find some of this activity painfully obvious, and others may require a bromide to help it go down. There should be enough that is both quasi-independent of these troublesome portions and palatable, however, so that the loss will not be significant if the reader wishes to bypass some of my mathematical enticements.

I describe some sections as "quasi-independent" of others because I have intentionally spread a number of the themes throughout the book rather than attempted to be exhaustive in each section. Thus, for example, I have chosen the concept of genius in chapter 2 to attempt to capture the centrality of problem solving in the culture of mathematics (in "Centrality of Problem Solving in Mathematics"). The centrality of problem solving in this spirit is first elaborated upon and then tempered considerably in chapter 4 in the section entitled "Problem/Situation Reverberation: A Return to Progress in Mathematics." There I point out how progress in the discipline of mathematics is sometimes achieved not by our standard notion of solutions to problems, but rather by reconceptualizing a field—something which is quite different from problem solving as depicted in the model of genius. In chapter 6 in the section "Thinking Like a Mathematician: A Critique," I return once more to the issue of mathematical talent and explore the tension between the genius notion of mathematical progress and the humanistic educational agenda that has been proposed in this book. Each elaboration/extension draws upon new machinery and/or worldviews that have been developed further in the appropriate section.

Other important themes recur throughout the text for which there may be minimal signaling to advertise their importance. One such theme has been that of self-referentiality, mentioned explicitly as a subsection only once—in connection with humor. Another is that of constructivism. The decision to have such themes pervade the text rather than set them off in sections of their own was a difficult one. They have been central (frequently implicitly) to so much of the text that I decided not to wrap them up, but rather have them operate as "zeigarniks" throughout.[3] While sympathetic with much of the orientation of constructivism, and while there has been a considerable amount of scholarship associated with it, it is a theme that has been so bandied about and uncritically incorporated within educational jargon, that I did not want to feature it without a more full-blown critical analysis, and to do so would have resulted in quite a different book and one of considerably greater length. I have, however, "sprinkled" issues related to constructivism throughout the book as they impinge upon themes of problem posing, problem solving, understanding, and especially meaning and its ambiguous uses. In particular, I will be pointing out the powerful legacy left by Dewey in establishing some of the essential and controversial aspects of the movement—a legacy that predated the present agenda by more than three-quarters of a century. I see the other theme, self-referentiality, as a largely unexplored jewel, and I wanted to organize my thinking about it so that the reader would appreciate its unexpected appearance as a force that could rejuvenate concepts that might otherwise be seen as "old hat."

Relation to *The Standards*

In reading this book, it is important to appreciate the sense in which my perspective relates to *The Standards* and associated documents. Though the talk I gave at the Philosophy of Education Research Center was focused heavily upon those documents, and though I shall discuss in general terms an array of concerns expressed in *The Standards*, this book is not intended primarily to be a commentary on those documents per se.[4] In focusing on two of the many themes that have become associated with them, I eventually came to use them as a backdrop to enable me to work through my own paradigm shift in mathematics education. As such, the issues that I raise transcend the bounds of those documents. Such issues include the misguided conception of the educational potential of problems, the relationships between problem posing and solving, the revealing of "self" in connection with others, and the value of thinking of "real-world" applications in a more humanistic way than is found in most of those documents.

As indicated above, some of what I reflect upon has personal roots that predate the publication of *The Standards* by more than two decades, and I will be integrating some of my earlier work with my more recent analysis. Interestingly enough, some of my criticism of the "new math" movement applies as much to the more recent reform of *The Standards* as well. Once again, my commentary on the phrase "thinking like a mathematician" is a case in point.

I have mentioned that this book represents something of a paradigm shift. As with early stages of such shifts, there are remnants of the old worldview that are left intact. Some of this is intentional. Some may be unnoticed or unacknowledged. I appreciate that I am straddling a number of different worldviews, including competing philosophical orientations, and that resolution may not only be difficult, but premature and inappropriate as well.

There are places, for example, at which I downplay concerns with mathematics as a form of propositional knowledge. I have done so with the intention of encouraging educators to carry on conversations with students in a different voice—one that respects issues for which "right and wrong" do not apply with ease. At other times I make use of a more conventional view of mathematical thinking in order to provide a backdrop against which students can view other aspects of their lives. I do this for example, in seeking connections between mathematics and humor, and also in reexamining some neglected areas of critical thinking (as in the concept of "same/different"). Thus, conventional views of mathematics are sometimes taken for granted and sometimes challenged, depending upon the educational purposes I wish to explore. In some places, I will be focusing upon the processes of coming to know, understand, doubt, and wondering at mathematical ideas. In others, I will ask the reader to focus upon a set of conclusions with minimal concern for how they were acquired, but maximal concern for the broad educational implications of such conclusions. The discussion of three different problems with regard to the concept of "with regard to" in chapter 6 is a case in point.

Perhaps a clearer way of highlighting the different roles portrayed by the set of mathematical experiences in this book is to distinguish among problems (and situations), solutions, and reflections on them—both as separate entities and in relation to each other. Apple (1992), for example, in calling for problems that are more closely aligned with student experience would seek examples that are politically or economically relevant, such as health care, nutrition, or job loss. While that is surely a reasonable tactic, it is also possible to choose problems whose *content* appears to be rather conventional and even whose solutions appear to be straightforward, but to reflect upon these problems and their solutions in a way that reveals matters that are far from irrelevant and impersonal. The problems of the census taker and the three women sitting in a circle in chapter 5, for example, appear to be variations of problems or puzzles that lack any sort of personal import. Reflection on solutions to these problems, however, engages us in a conversation not about *content* but about *style of thinking* that connects the mathematical experience with issues of morality. Thus, one should be cautious about dismissing content per se as "old hat." What has the earmarks of traditional curriculum will on numerous occasions achieve unexpected rejuvenation as we reflect upon problems, situations, and purported solutions.

Acknowledgments

I am grateful to a number of people and institutions for allowing me the time and peace of mind to devote to this enterprise. Hugh Petrie and Nancy Broderick, Dean and Associate Dean of the Graduate School of Education at the University at Buffalo at the time, were thoughtful in orchestrating a year's leave from my home institution. I also wish to thank Israel Scheffler for sponsoring my participation at the Philosophy of Education Research Center. Colleagues at PERC influenced my thinking about many matters by their broad, diverse, and stimulating dialogue on issues of education and society.

I am also grateful to the National Science Foundation for funding projects that influenced this work. In collaboration with Thomas Cooney, my initial NSF project focused on implicit beliefs held by teachers regarding the nature of education and of mathematics. Together with a number of other colleagues, we subsequently made use of some of those findings to design teacher education experiences which expressed the views of *The Standards*. It was at the tail end of my work on that project that I assumed a more detached stance toward *The Standards*—a stance that provided the inspiration for this book.

Every member of Peter Lang Publishing, Inc. with whom I have interacted has been most considerate and helpful. Heidi Burns, who did the copyediting, used a magnifying glass of enormous power in order to fashion a more readable text. From the moment the manuscript was accepted, Phyllis Korper—acquisitions editor—as well as her assistants responded with kindness and wisdom to my ongoing concerns. She consistently reminded me that matters of substance, organization, and style were ultimately my decision regardless of excellent suggestions for revision. I appreciate her attentiveness more than I think she realized. Lisa Dillon, and her staff, who were responsible for production, graciously assumed numerous formatting tasks. If left to my own devices, the publication of this book would have been delayed until the next revolution in mathematics education.

I thus complete a project that has been in various stages of gestation for several years. Or has it been several decades? To the extent that it was the latter, I am grateful to my students and colleagues from a variety of institutions—including Harvard Graduate School of Education, Syracuse University, Hebrew University in Jerusalem, the University of Georgia and the State University of New York at Buffalo—my home for a quarter of a century. I have learned more than I can express from my own children (now grown), Jordan and Sharon. Many of these people appear as characters in one form or another in the drama that unfolds.

My greatest debt is to my wife, Eileen, not only for urging me to extend my brief talk into something a bit more engaging, but also for encouraging me to absent myself for long periods of time from her addictive companionship. I did so by going into my third den, a reconstructed closet—just as others were coming out of theirs.

Finally, I would like to acknowledge the following publishers for their kind permission to reproduce the material indicated below:

> For the Learning of Mathematics Publishing Association, Kingston, Ontario for permission to include the following from the journal, *For the Learning of Mathematics*: (i) excerpts and a diagram from Brown, Stephen I. (1981), Ye shall be known by your generations, *1*(3), 27–36; (ii) excerpts from Brown, Stephen I. (1984), The logic of problem generation: From morality and solving to de-posing and rebellion, *4* (1), 9–20; (iii) excerpts and a diagram from Borasi, Raffaella, & Brown, Stephen I. (1985), A novel approach to texts. *5*(1), 21–23. (permission granted from Raffaella Borasi as well).
>
> Heinemann, a division of Reed Elsevier Inc. Portsmouth, NH, and John Dossey for permission to include (i) excerpts from Brown, Stephen I. (1996), *Posing mathematically*; (ii) figure 5.1 from Dossey, John (1996), *Modeling with functions*.
>
> Kluwer Academic Publishers. Dortrecht, The Netherlands for permission to include excerpts from Brown, Stephen I. (1996). Towards humanistic mathematics education, (pp. 1289–1321). In Alan Bishop, Ken Clements, Jeremy Kilpatrick, Colette Laborde & Christine Keitel. (Eds.), *International handbook in mathematics education*.
>
> National Council of Teachers of Mathematics, Reston, VA, for permission to include a portion of the text and accompanying diagram from p. 138 of *Curriculum and evaluation standards for school mathematics* (1989).
>
> The Yale Law Journal Company and Fred B. Rothman & Company, New Haven, CT, for permission to include a modified version of the figure on p. 1842 from Lukinsky, Joseph (1987), Law in education: A reminiscence with some footnotes to Robert Cover's *Nomos* and *Narrative*. *96*(8), 1836–1859.

Lastly, and in honor of the emerging "post-finally" era, I would like to single out six colleagues, each of whom has had a powerful influence on my intellectual and interpersonal development for at least two decades. My collaboration with Marion Walter began when we were both on the faculty of the Harvard Graduate School of Education. I met Thomas Cooney during a sabbatical leave at the University of Georgia. My work with Joseph Lukinsky began with our co-teaching courses that were cross registered between each of our home institutions—Harvard and Brandeis University. Three other members of the two-decade club with whom I collaborated in one form or other during my stay at the University at Buffalo are Mary Finn, David Nyberg and Gerald Rising.

Part I

BACKDROP AND PERSPECTIVE

Chapter 1

Reform in School Mathematics: A Backdrop and Critical Overview

> To my knowledge, Socrates infrequently answered a question and never solved a problem. . . . Neither Plato nor Socrates ever assumed that the accomplishments of Greek scientists reflected a model of question asking and problem solving that was appropriate to the development of the good society.
> —Seymour B. Sarason

"Mathematics education" became less of an oxymoron during the last decades of the twentieth century than ever before. A major contributing factor has been *The Standards* movement which shares an important feature of the "new math" of the generation that preceded it: an appreciation for the dysfunctionality of mathematics perceived as a collection of rules to be learned in rote fashion. It outstrips its predecessor, however, in its awareness of and concern with educational matters *writ large*.

The curriculum of the late 1950s and early 1960s attempted to confront a popular view of mathematics as a concatenation of an unmotivated, unrelated, and arbitrary set of rules. Instead an effort was made to create curriculum that would reveal the structural nature of the subject, at least at the secondary-school level. The goal frequently touted was to involve students in "thinking like mathematicians"—a catch phrase that placed heavy emphasis upon reproducing established bodies of knowledge.[1] Students were encouraged to "discover" truths in gentle and inductive ways. Some of the truths were selected as axioms, and others would eventually be proven by reliance upon a relatively small and coherent collection of axioms. This point of view was fostered not only in geometry (terrain within which deductive proof was a popularly accepted concept), but it became the mainstay in algebra and other fields that had been thought by nonmathematians to be the province of arbitrary rules.

Below is an excerpt from a "discovery" exercise from University of Illinois Committee on School Mathematics (1961),that was designed to introduce a basic axiom of algebra—the distributive property: $a \cdot (b + c) = a \cdot b + a \cdot c$ [depicted here is: $(a \cdot b) + (c \cdot b) = (a + c) \cdot b$].

> Perhaps you have found short-cuts for some problems which involve both multiplication and addition.
> $$7 \cdot 11 + 3 \cdot 11 = ?$$
> Do you see a short way of solving this problem? If you don't, you may see it after you have filled in the blanks in the following sentences:
> $4 \cdot 15 + 6 \cdot 15 = \underline{60} + \underline{90} = 150 = \underline{10} \cdot 15$
> $8 \cdot 29 + 2 \cdot 29 = \underline{} + \underline{} = \underline{} = \underline{} \cdot 29$
> $13 \cdot 21 + 17 \cdot 21 = \underline{} + \underline{} = \underline{} \cdot 21$ (Vol. 1, p. 50)

What is particularly clever about this example—and many other exercises like it—is that it is crafted in such a way that there is enormous immediate payoff in understanding the principle. Specifically, this principle provides a short-cut for doing mentally what otherwise might involve laborious calculation (especially in the absence of calculating devices). That is, while $(8 \cdot 29)$ and $(2 \cdot 29)$ are calculations that would not normally be derived mentally (without using this very principle at least implicitly in arriving at each of the two products), the calculation of $10 \cdot 29$ can be completed without aid of calculator or pencil and paper. Using properties of the above sort—in combination with each other—allows for the logical derivation of other principles and properties that otherwise were taken as a collection of arbitrary rules and regulations.

What emerged subsequently, from the mid 1980s through the 1990s, however was a plea for curriculum reconstruction in mathematics that was a less linear, less elite view of the subject than was the case with the "new math." It is a view that invites integration with other fields and other experiences and is a testimonial to critical voices in the era of the "new math" that had been strongly resisted.[2] While the curriculum of the "new math" seemed to further exacerbate the differences between "talented" and "slower" students, its successor responded to a more egalitarian impulse.

The new reform movement of the 1990s, associated with a focus on national standards, expresses deep pedagogical concern not only by professional mathematics educators, but by mathematicians and scientists as well.[3] While these latter groups also influenced the movement of the fifties and sixties, at that time their concern with pedagogy was primarily instrumental—designed for the purpose of portraying a more scholarly view of the discipline. As such, it was concerned with cleansing ambiguous language (sometimes caricatured in such distinctions as that between "number" and "numeral") and with replacing essentially meaningless manipulation of symbols with a greater appreciation for proof and for the logical status of assertions.

The transition towards an egalitarian rather than an elitist view of the discipline with an implied concern for pedagogy is summarized by Lynn Steen (1990b)—a mathematician and former president of the Mathematical Association of America. He comments:

> The first steps in bringing about change must be to convince the public of certain realities:
>
> - That mathematics is the foundation discipline for science and technology.
> - That far too many minority children leave school without having acquired the mathematical power for productive lives.
> - That all children—not only those with special talents—can learn mathematics.
> - That our children must learn a different kind of mathematics for the future from what was adequate in the past.
> - That confidence rather than calculation should be a chief objective of school mathematics.
> - That our nation's economic future depends on strength in mathematics education. (p. 134)

The above summary of course does not itself a pedagogy make, nor does it disclose many concerns found elsewhere in the reform of *The Standards*, but it is the beginning of an argument that focuses on the need to reexplore what and how we teach students of all abilities.

Both the current reform and the earlier "new math" were in part motivated by a perceived loss of our competitive international edge. While the Soviet launch of Sputnik into outer space was a major force that defined our inferiority in the era of the "new math," *The Standards* movement is driven by a form of competition as well: concern with low mathematical performance on international tests of achievement. Though it may be possible to reduce that concern ultimately to economic roots, *The Standards* documents are more concerned with internal issues that are directed towards the disparity in performance along racial and social class lines.

A Problem Solving Perspective

The centerpiece for the new reform in school mathematics is problem solving. It is the glue that pervades and joins the other dimensions: the multiple conceptions and roles of reason in mathematical thinking;

connectedness within mathematics and between it and other fields; and mathematics as an act of communication. The point of view is summarized in the *Curriculum and Evaluation Standards* (National Council of Teachers of Mathematics, 1989):

> Mathematical problem solving, in its broadest sense is nearly synonymous with doing mathematics. Thus whereas it is useful to differentiate among conceptual, procedural and problem solving goals for students in the early stage of mathematical learning, these distinctions should begin to blur as students mature mathematically. In grades 9-12, the problem solving strategies learned in earlier grades should have become increasingly internalized and integrated to form a broad basis for the student's approach to doing mathematics, regardless of the topic at hand. From this perspective, problem solving is much more than applying specific techniques to the solution of classes of word problems. It is a process by which the fabric of mathematics as identified in later standards is both constructed and reinforced. (p. 137)

It is clear from the above commitment that problem solving is as much concerned with process as it is with product. It is largely motivated by a desire to encourage people to confront intelligently a creature that is "wild" rather than to reproduce a set of specific techniques guaranteed to further "tame" an already well trained beast. It is as much concerned with students learning heuristics for problem solving and applying these strategies to deal with unknown entities as it is with having them reproduce minor variations on what is already well known.

Below are some typical problems that are proposed for grades 5 to 8 in this document:

> Maria used her calculator to explore this problem: Select five digits. Use the five digits to form a two digit and a three digit number so that their product is the largest possible. Then find the arrangement that gives the smallest product. (p. 76)

> How many handshakes will occur at a party if every one of the 15 guests shakes hands with each of the others? (p. 77)

In both cases, it is assumed that the students have not previously solved problems of the "type" selected. They are encouraged to seek not one but multiple ways of understanding the problems. Trial and error thinking, among other strategies, is given its due. In the first problem, just selecting many two-digit and three-digit numbers at random might be a reasonable start. Creating a table of this information would be helpful. Noticing any commonalties among the array of products in relation to the two-digit and three-digit numbers selected could lead to some interesting patterns to explore. In the handshake problem, drawing a sketch might enable students to understand what the problem is "asking." Perhaps selecting a variety of similar cases to the one proposed and attempting to graph relationships (for example, the number of handshakes as a function of the number of people shaking) would help. Although the second problem might be solved easily by students already familiar with combinations and permutations, it is not a task students can approach formulaically in the grades for which it is intended in *The Standards*.

Rooted in a conception of thinking compatible with Dewey's (1910) "reflective thought," mathematicians and educators borrow from and expand upon a model of problem solving that is associated with the work of the world renown mathematician/problem solver, Georg Polya (1954, 1957, 1962). Although they need not necessarily be applied in strict sequential order, and although they intertwine with each other, essential elements of Polya's model are:

1. gain an awareness or understanding of the problem,
2. consider possible strategies for solving it,
3. choose a strategy,
4. carry out the strategy
5. verify the solution
6. look back to see what has been learned.

In addition to focusing on strategies per se, there is an appreciation for partial or approximate solutions to problems rather than for complete and precise ones. There is also an increased appreciation for the use of multiple means of characterizing specific problems, including the use of concrete materials, graphing strategies, and intertwining algebraic and geometric perspectives.

The incorporation of the handheld calculator and the computer as tools rather than ends in themselves further supports problem solving as an alternative to mindless calculation for mathematical experiences at all levels. The fact that such devices can do simple arithmetical operations at the grade school level, can perform algebraic simplification in the secondary school curriculum, and can integrate and differentiate functions in the calculus all become an invitation to seek alternatives to learning algorithms as a major component of the curriculum.

Before offering a critical perspective, we turn to a second major focus of the book: applications of mathematics and its connectedness to other experiences.

Applications

The following problem, taken from the National Council of Teachers of Mathematics' (1989) *Curriculum and Evaluation Standards for School Mathematics,* depicts the applications orientation of the reform movement.

> Suppose Anne tells you that under her old method of shooting free throws in basketball, her average was 60%. Using a new method of shooting, she scored 9 out of her first 10 throws. Should she conclude that the new method really is better than the old method? (p. 172)

Dealing with uncertainty and conveying an implicit nonsexism, the problem encourages students to apply what they know about probability to a novel "real-life" type of situation.

Problem solving of this sort has many advantages, especially over standard word problems that appear to be contrived and that do not connect with "real-world" experiences (see Nickson, 2000). Encouraging such connections not only has the potential to motivate exploration of uncharted territory, but in addition it encourages students to better understand the mathematical ideas behind the exploration.

The variety of connections proposed for use in curriculum is impressive. Below is a list of suggestions from the same document:

- *Art:* the use of symmetry, perspective, spatial representations, and patterns (including fractals) to create original artistic works.
- *Biology:* the use of scaling to identify limiting factors on the growth of various organisms.
- *Business:* the optimization of a communication network.
- *Industrial arts:* the use of mathematics-based computer-aided design in producing scale drawings or models of three-dimensional objects such as houses.
- *Medicine:* modeling an inoculation plan to eliminate an infectious disease.
- *Physics:* the use of vectors to address problems involving forces.
- *Social science:* the use of statistical techniques in predicting and analyzing election results. (p. 149)

It is clear that the intention of such documents as *The Standards* is to integrate the focus on problem solving with that of applications. They are both part of a cloth that encourages students to confront an epistemological attitude that permits ignorance to become an excuse for not participating in inquiry. Every effort is made to enable students to connect what they do not know with what they already know. Use of concrete materials, special cases, data, graphs, collaboration with others, and deductive proof when it seems appropriate are all part of the machinery that is included in such reform.

Not only are problems now included in the curriculum that are more enticing than the aforementioned word problems associated with algebra in an earlier era (distance, work, coin, mixture), but some of them are a bit "messier" than the problem of Anne and her basketball endeavors. Below are excerpts from two essays that appear in the journal *The Mathematics Teacher* in the late 1990s.

> "Do I have to know how to do any math for geography?" Although Luke was hoping to find a niche where he needed as little mathematics as possible, his question brought to mind all the practical mathematics that I had used in college geography projects. I was motivated to develop an extended hands-on project for my eighth-grade mathematics classes in which they were to determine the probability that a local creek would flood. This project can be easily adapted to almost any secondary-level mathematics class. Input and support from core-subject teachers on our academic team led to the development of a complete interdisciplinary unit that incorporated technology and encouraged writing across the curriculum. The project was assigned to all students on the academic team, whose skill levels ranged from remedial to honors abilities. (Haug, 1998, p. 456)

> Students are aware that germs spread disease. They also know, at least on an intellectual level, that they can avoid catching some diseases by avoiding risky encounters with infected individuals. The definition of risky encounter varies with the illness. For example, such illnesses as the common cold may be spread by an activity as common as shaking hands, whereas AIDS is frequently spread by sexual contact. Cures do not yet exist for either of these illnesses. The goal of this activity is to model the exponential growth of the common cold, AIDS, or any other communicable disease. (Stor & Briggs, 1998, p. 464)

Unlike the basketball example, the problem of determining whether or not the local creek would overflow does require expertise from other disciplines in order to find out what relevant variables might be isolated and subjected to mathematical analysis.

Both of these examples make explicit an element that was implicit in the example of Anne. In the case of the potentially flooding creek, they had to determine what was relevant about the real world and what was "noise" for the purpose of creating a mathematical design. This distinction required minimal attention in the Anne example, for though the fact that Anne was female has a political correctness to it, it is obvious (at least as the problem is framed) that her gender does not affect the mathematical analysis.

The example of the spread of disease makes the point of mathematical design based upon relevance vs. noise even more explicit. That is, instead of exploring the actual passage of disease by having students infect each other, the teachers created a highly imaginative unit that involved a form of simulation. They made use of dice, and as pairs of students interacted, each would role a die. If the sum was five or less, then the encounter was defined as "risky."

Connecting mathematics to the real world is an invitation to introduce two fundamental concepts: that of model and that of function. Some mathematical model is developed in such a way that salient objects of the real world are associated with mathematical elements, and activities in the real world are associated with mathematical operations. Frequently the functional relationships among variables in the mathematical model are explored in such a way that the results of their connections can then be used to understand or predict something about the real world from which the model was derived. They are powerful concepts which enable students to understand that mathematics and the real world are essentially two different systems that are being linked and coordinated in such a way that the known behavior of one system has the potential to shed light on the lesser known behavior of the other. In its most ideal form, the mathematical model is isomorphic to those aspects of the real world that have been isolated and retained.

A Critical Perspective

This program acknowledges not only the diversity within and among fields, but it also is sensitive to the fact that learning is an active rather than a passive experience. In addition, there is an awareness of the social context within which learning flourishes.

Focusing on learning as the construction of meaning and on the social arena that influences what is learned, the movement reflects elements of a theoretical orientation associated with constructivism and interactional thinking. Both have roots that go back to the first quarter of the twentieth century, and although the language may have changed, it is clear that the movement has been influenced by the philosophical/psychological perspectives of scholars such as Dewey, Piaget, Vygotsky, and others.

Criticism of these documents comes from a variety of points of view. The fact that the introduction of national standards in mathematics was the first of over a dozen documents that were slated to appear in other curriculum areas made their scrutiny that much more compelling. Some of the criticism is based upon a perceived lack of educational research that might predict student success in such programs, e.g., Gardner (1998). From a more conceptual perspective, there is significant debate over whether or not problem solving strategies are generalizable among fields or even within subfields of a given discipline. This question is of particular concern among advocates of programs in critical thinking as well as problem solving. Schoenfeld (1980) was one of the first to raise questions about the context specific nature of problem solving heuristics in mathematics. McPeck (1990) raises the issue with regard to critical thinking and subject matter in general, and there has been a virulent debate within the field of philosophy of education for the past few decades over the value of isolating general critical thinking strategies that do not derive from a deep knowledge of specific fields of inquiry.

Some critics have commented on the culture of schools and wonder about the relationship between the education of teachers or the authoritarian nature of the school's organization and the open inquiry expected in classrooms designed around such curriculum—a reassertion of Sarason's (1971) earlier criticism of the "new" math of the previous generation. This perspective has been elaborated upon by some critics who were concerned that such an inquiry oriented curriculum achieved its goals without adequate attention to subtle social class distinctions. Those students who come from environments that already honor dialogue and critical inquiry would be privileged while others might be alienated by such an approach to the curriculum. Concerned with the reality rather than the rhetoric of "mathematics for everybody," Apple (1992) points out that unless teachers are educated as social critics, they will lack an appreciation for the subtle forces of a mathematics curriculum that produces precisely the opposite of what is desired. He fears that despite rhetoric to the contrary, the democratic impulse of the reform movement will be taken over by a more conservative ideology.

Some scholars interested in the culture of schools have had more of an external than an internal focus. Appelbaum (1995), for example, has examined such matters as the life of the classroom in relation to the education of teachers or of school organization from the perspective of power and identity as expressed in public discourse and popular culture. Using the case of Jamie Escalante (mathematics teacher "in the barrio" in a Los Angeles high school), for example, he explores how it is that popular culture has come to define "super teacher" in highly individualistic terms—thus "making 'inconsequential' institutional and social contexts of teaching" (p. 66).

While some have questioned the clarity and mathematical significance of various standards that have been proposed (Raimi & Braden, 1998; MacLane, 1998), others have debated the appropriateness of national standards in a democratic society. That is, what role should local rather than national communities play not only in implementing, but in defining what and how their children should be taught?

Interestingly enough, although communities might differ, there is a powerful ideological stance that affirms the value of everyone acquiring some significant body of mathematical knowledge, if for no other reason than as a ticket for entrance into many careers. This point of view is expressed most clearly in *Everybody Counts* (National Research Council, 1989). Noddings (1994) questions why educators and the public at large accede to such an ideology, and she offers alternative perspectives.

As important as these issues are, the orientation in this book is of a different sort. My intent is to widen educational and curriculum lenses in the areas of problem solving and real-world connections. Issues of curriculum design are presented in the language of educational possibility rather than program implementation. There is a considerable need for elaboration, clarification, and modification of what I will be criticizing in this book. In addition, an essential next step is to incorporate some of the above mentioned criticism with my own agenda in order to create a more viable curriculum.

Before turning to my own critical perspective, it is necessary to appreciate that there are in fact some elements within existing curriculum that exhibit pieces of the wider angle lens that I am attempting to craft. For the most part, however, they have not been incorporated with as full an awareness of what they are resisting and with an inadequate appreciation for their potential to reconstruct the educational scene. Though some critical commentary will be illustrated by example, the focus here is more on seeking a sound rationale for such reconstruction than on documenting pockets of excellence.

In an attempt to define the first of the two major concerns of this book, I would like to suggest that the reader participate in a variation of a popular television show—the game of jeopardy. Have patience. Allow yourself to experience the impact of the game before trying to locate the punch line. It would not be a travesty for ganders to eat firsthand what they advocate as feed for the geese.

Exposing the First Theme:
A Variation on the Game of Jeopardy

The game of jeopardy inverts the normally accepted sequence of question and answer. In that game, we are given an overriding category together with an answer and are expected to come up with "the" question. Despite the fact that a supposed answer has any number of equivalent and nonequivalent questions associated with it, our common experiences and especially our test-taking acumen frequently enable us to predict and agree upon an associated question for some answers. At least that is the case in the game of jeopardy as it is played on television. Some simple examples would be:

- Category: **Sports**
  [Answer]: The team that appeared four consecutive years in the super bowl and never won.
  [Question]: Who are the Buffalo Bills?

- Category: **Famous Ships**
  [Answer]: The Pinta, the Niña, and the Santa Maria.
  [Question]: What are the names of the three ships that Columbus set sail on from Spain?

Surely, the above questions for each answer are not unique from a strictly logical point of view, even if empirical evidence suggests that most people confronted with a specific answer come up with the same question. A question associated with the three ships as an answer, for example, might be: "What are three ships whose names begin with the letters "P," "N," and "S"? In addition, there are some answers that would not work well as starting points in the game of jeopardy. For example, the answer "red," or the answer "four" are answers that most likely would not conjure up a single question that can be predicted by a majority of people regardless of their specific knowledge and regardless of the overriding category. While it would be interesting to figure out the requirements of answers that seem to fit well into a game of this sort, let us bypass that issue in favor of playing a variation of the game.

The variation is as follows: Instead of starting with one answer, let us be more generous and provide a set of answers for a category. Here are two sets of answers (X the first, and Y the second) both belonging to the overriding category of mathematics education. What question do you think is associated with each set?

X
First Set of Answers

1. Solving problems gets you to think.
2. Problem solving is what math is all about.
3. Interesting problems enable you to practice basics in a disguised way.

Y
Second Set of Answers

1. If you learn to solve problems well in mathematics, you will be better able to solve problems in your life.

2. Problem solving is a good way of capturing students' interest in something they might otherwise not want to think about.
3. Reflecting on problems you have tried to solve, gets you to think about your thinking process.
4. Working on problems is one of the most human activities. It is what separates us from lower organisms.

To set the stage for my forthcoming criticism of problem solving in *The Standards*, I urge that you try to figure out what question might have generated each of the two sets of answers described above. *Please do so before reading on.*

Actually, this is an activity I used with a group of experienced mathematics teachers. Each student was asked to guess the question that had generated each of the above two lists. Some of the questions that were suggested for each of these sets of answers are listed below.

Questions Suggested for the First Set:
1. Why do we solve problems in mathematics?
2. Why is problem solving effective in the classroom?
3. Why is it important that we learn how to solve problems?

Questions Suggested for the Second Set:
1. What are the benefits of learning problem solving techniques?
2. How is problem solving educational?
3. Why is it important to include problem solving in the curriculum?

Behind This Game of Jeopardy

Interestingly enough, the two "answer" sets (**X** and **Y** above) look quite similar. They all speak about or imply that problem solving is the topic for which answers have been supplied. It is difficult to imagine that these answer sets might have come from different questions. It is now time to reveal how the original two sets (**X** and **Y**) were generated. There is actually an empirical basis for them described below.

A few weeks before presenting this activity to my students, I had given a talk to a philosophy colloquium at a local college. I began the talk by passing out a piece of paper to each member of the audience, and asked them all to respond in writing to the question that was printed on the top of each paper. Though everyone thought that the same question was being asked, in fact I had created two different questions. Half the audience received one question (**A** below) and the other half received another (**B** below). Thus, in half the cases, they responded to the following:

A

What are some good reasons for including problems in the school curriculum?

The other half, responded to a different question:

B

What are some good reasons for including problem solving in the school curriculum?

A quick reading might suggest that the two questions (**A** and **B**) are literally the same. It would probably be worthwhile rereading them in order to catch the difference.

As I called on volunteers in the philosophy colloquium, I placed answers **X** from those who were responding to the first question, **A**, on one overlay, while I placed answers **Y** to the second question, **B**, on a different overlay. Not knowing that they had been asked different questions, people in the audience then looked at the two different overlays (**X** and **Y**) and were asked to let us know what similarities and differences they saw in the pattern of responses between both sets—each taken as a whole.

Although they did notice that there were some points made on one overlay that had not been made on the other, in an audience of over fifty people, they saw the two overlays as being driven by the same concern. That is, they all concluded that it must have been the same question that generated both lists **X** and **Y**.

It might be enlightening for the reader to test this activity with different audiences. I would suggest presenting it as I did in the case of the philosophy colloquium. The jeopardy format I used to introduce this section (with the "answers" already given) was done as a necessary variation due to the fact that we (the reader and myself) were not interacting in a direct way, and I did not want to reveal my point about the tendency to reduce "problem" to "problem solving" until there was an opportunity for the reader to personally experience that inclination—one that took many years for me to identify and to disentangle.

A large portion of this book is dedicated to an analysis of the reason for why it is that we have this inclination, to a consideration of the educational shortcomings of this connection, and to alternative perspectives. I am driven by the realization that problems have educational potential and educational uses considerably beyond (and even distinct from) their connections with solutions or with efforts to come up with solutions. A wonderful irony of appreciating that problems may have lives that are independent of their solutions is that once realized, it is possible to revisit the concept of problem solving itself with fresh insights.

The Second Theme: Real World and Mathematics

This section requires less patience than our introduction to the first theme. My reexamination of the relationship between mathematics and the "real world" derives from a different impulse from that of the problem solving theme. It challenges the notion that the connection between mathematics and other domains is necessarily one that can be characterized via the notion of a model—at least model in the sense that we have characterized it so far.

There is nothing inherently wrong with the notion of a model as far as it goes. In fact, much progress in the "real world" is achieved by so translating real-world problems. What is missing from an educational perspective, however, is that such a notion of application tends to isolate mathematics. That is, mathematics is frequently only a partial solution to real-world problems. Frequently one can go only so far with a mathematical analysis, and then it is necessary to introduce other dimensions of human thought and experience in order to think more about the problem.

What is educationally objectionable about the concept of a model (frequently used implicitly) as a dominant tool to integrate mathematics with the real world is not only that the incompleteness of mathematical thinking for the purposes of solving real-world problems is left hidden, but the nature of mathematical thinking itself is truncated by such an approach.

The above mentioned array of possible connecting categories from art to biology to social science leaves important elements of the humanities out. No suggestion is made that mathematics might relate to poetry. Here, I have in mind something quite different from the kind of contribution Edna St. Vincent Millay made in her famous poem, *Euclid Alone Has Looked on Beauty*. How is it, for example, that metaphor and imagery function in the way individuals remember, create, or convey to others an array of mathematical ideas?

Another example that reveals something of the nature of mathematics beyond its functioning as a model is to inquire into how mathematical thinking compares with humor. What are some similarities that are shared between mathematics and humor? Here we do not use one field of experience to solve a problem in another, but rather we are looking at the fields (and of course examples in them) in order to understand what the fields themselves are about.

It is not only that we will be exploring new fields in relation to mathematics, but as importantly, we will be revealing different conceptions of what it means to *apply* or to *connect* one field with another. In exploring alternative conceptions, we will also inquire into what it means for something to be real. What is *real* is something that requires a deeper analysis than we have allowed ourselves to imagine in the designing of curriculum.

The impulse that drives the profession to connect mathematics with the real world is frequently an ameliorative one. It is a way of approaching a field many students perceive not only as overly abstract but also as isolated from their lives. By connecting it with the real world—objects that supposedly can be touched and seen—we attempt to entice students to learn mathematics by centering on its relevance.

While such enticement is laudatory, it neglects to appreciate different levels of what is "real," and that some of them are not merely given at birth, but rather are subject to education. My wife, watching virtually any film, becomes like the character in Woody Allen's *Purple Rose of Cairo* (see Greenhut, 1985). She essentially enters into a different state of consciousness and that world becomes as real for her as anything in the here and now.

Mulling over what it takes to make us feel "more real," Nozick (1990) helps us understand what it is that accounts for this notion of reality even when the objects themselves have an existence that does not square with what we frequently take to be reality. He comments,

> We . . . are more real at some times than at others, more real in some modes than in others. People often say they feel most real when they are working with intense concentration and focus, with skills and capacities effectively brought into play; they feel most real when they feel most creative. Some say during sexual excitement, some say when they are alert and learning new things. We are more real when all our energies are focused, our attention riveted, when we are alert, functioning completely, utilizing our (valuable) powers. Focusing intensely brings us into sharper focus. (p. 131)

In laying out a vision of reality that is considerably more robust than the one presented to us for applying mathematics to the real world, Nozick comments on the limited vision of philosophers of mathematics when they myopically connect the concept of reality to that of "existence." He says,

> Mathematicians . . . delineate objects and structures wherein sharp properties interlock in a densely layered network of combinatorial possibilities, relations, and implications. To ask, "Do mathematical entities exist?"—the question put by philosophers of mathematics—does not capture the saliency of their vivid reality. The Greeks could not fail to be captivated by such objects and the intricate patterns they exhibited so definitely and sharply, even in the case of "irrational" numbers, which were incommensurate. Tradition reports that Plato held that Forms—according to his theories, the most real entities—were (like) numbers. The mathematical realm, a vivid one, grips our attention because it is so real. (pp. 130–31)

Anyone who has observed young children between the ages of three and about nine years of age knows that they are thoroughly captivated and live in a world inhabited by imaginary creatures—dinosaurs like Barney, and knights and princes and princesses. They are at a stage that Egan (1979) characterizes as "mythic." What frequently transpires in the education of youngsters is that they are taught to outgrow that sense of reality and learn to replace it with something more mundane and supposedly more "grown-up." Ashley Montagu (1981) has much to say about the sense in which this view inverts what is the natural evolution of our closest relatives in the animal kingdom. He makes a strong case for enabling humans to "grow young." In discussing the "real world" and its relation to mathematics, we will be inviting the reader to entertain a more robust view of reality—one that is consistent with what Montagu describes as the science of neoteny.

A Humanistic Perspective

In this section, we sketch a perspective that directs the way for the reconstruction of both problem solving and real-world connections. That is, it is one thing to identify an essential fallacy in collapsing problem and problem solving for educational purposes or in exposing a limited and misguided view of what is "real." It is another to offer a vision of what might replace or enlarge what exists. I view a humanistic orientation not so much as a set of schemes and activities, nor even as a philosophy but more as a way of seeing the world. It points in directions that are not usually an explicit part of the mathematics curriculum. It does not necessarily tell us what to do in specific circumstances, but rather entices us to think differently. Many of the ideas sketched in this section will be elaborated as we apply them throughout this text.

The humanistic conception has a long history and is one that has conflicting schools of thought.[4] We shall take as its essential ingredient a more explicit vision of human agency in portraying the nature of mathematical thought and in teaching it as well. We introduce the theme by citing two commentaries that represent negative reactions to the "new math" rather than to the reform associated with *The Standards*. Though *The Standards* do respond positively to some of the criticism implied by these anecdotes, there are numerous remnants that remain to be excised.

The first is by an articulate woman who participated in a study conducted by Buerk (1982). These women were outstanding in some intellectual field but had inordinate fear of mathematics at any level.

And on the eighth day, God created mathematics. He took stainless steel, and he rolled it out thin, and he made it into a fence forty cubits high, and infinite cubits long. And on the fence, in fair capitals, he did print rules, theorems axioms and pointed reminders. "Invert and multiply." "The square on the hypotenuse is three decibels louder than one hand clapping." "Always do what's in the parentheses first." And when he finished, he said "On one side of the fence will reside those who are bad at math, and woe unto them, for they shall weep and gnash their teeth."

Math does make me think of a stainless steel wall—hard, cold, smooth, offering no handhold; all it does is glint back at me. Edge up to it, put your nose against it; it doesn't give anything back; you can't put a dent in it; it doesn't take your shape; it doesn't have any smell; all it does is make your nose cold. I like the shine of it—it does look smart, intelligent in an icy way. But I resent its cold impenetrability, its supercilious glare. (p. 19)

The second is by a college student who eventually majored in mathematics but who was influenced by the dehumanizing effect of a program that derived its pedagogy directly from its conception of mathematics exclusively as a logical, deductive field.

In my junior year at college, I took my first graduate level mathematics course—finite dimensional vector spaces. It was offered by a professor who had an international reputation. The first day, he told us that the only things that count in proving anything are axioms, definitions, rules of logic and previously established theorems. Any other crutch was to be interpreted as a bastardization of the discipline. He proceeded to list the axioms of a vector space, and as sometimes happens under such circumstances, he got stuck. He stood before us, mumbled a few inaudible words, and then turning his back to the class, and blocking the blackboard with a stomach that was adequate for the purpose, he sketched a tiny diagram that looked something like the figure below:

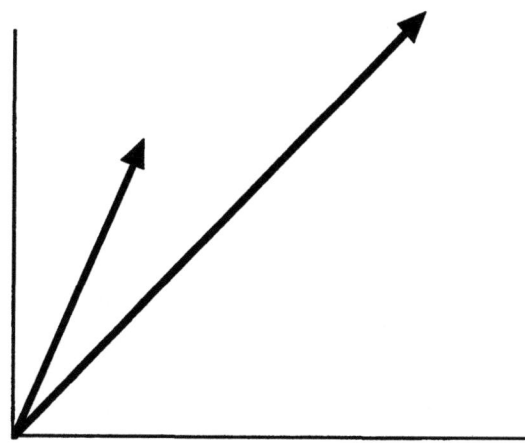

Attempting to be consistent with his original advice, he quickly erased his sketch and proceeded to list a few more axioms and to prove a few "baby theorems" based solely upon "axioms, definitions, and rules of logic."[5]

What do these anecdotes portray? They can be milked at length, but let us focus upon a few of their qualities in order to establish some of what will be eroded by a humanistic orientation.

1. The "Givenness" of It All:
   In both examples, the subject matter is presented as "given." In the first case, it is a perception about the entire field of mathematics; in the second it is about a particular field of inquiry. Much of this book will attempt to unearth what can be done to enhance and to challenge what is "given"—a point of view that applies as much to *The Standards* movement as to the "new math," though the nature of what is "given" may be different in the two cases. We will not only challenge the given for the purpose of understanding why it was selected, but also to see it as potential for thinking otherwise.
2. The Absence of "Self":
   In both cases, there is the basic assumption that the subject matter and the "self" are disjoint—and that neither has much to say about informing the other (except as an act of intimidation). We will inquire into what sort of encounters with subject matter has the potential to enable us to better understand how it is we think and feel.
3. The Domination of Logic:
   Both anecdotes convey the impression that logic alone rules how we are to make sense of the subject matter. There is no appreciation for the political, aesthetic, and social contexts that influenced its evolution, and that might direct its future development as well as the way in which it is learned. Furthermore, the basic ideas of the field are taken as having been born in the absence of labor pains and as uncontroversial. Even with the provision of experiential pedagogical activities, the field is frequently presented as the sine qua non of a form of inquiry whose conclusions are predetermined and considered to be certain.
4. The Mind as Hidden:
   What are we to make of the erased sketch of vectors in the above anecdote ? By hiding what he really thought about these axioms, we are led to believe that they held very little meaning for the instructor. If they were to be excluded for the purpose of proof, and if he was somewhat embarrassed by his need to resort to their use,

we are led to believe that it is a weakness to have to engage one's mind in creating images and metaphors that are fundamental human baggage in all other fields of inquiry.
5. The Isolation of It All:
We are led to believe that there is something unique about the way in which one thinks in mathematics. Though we might make connections for motivational purposes, and though we might tolerate or even encourage inductive strategies as part of the early stages in mathematical thinking, ultimately the deductive nature of mathematical thought separates it as a way of knowing from all other world experiences.
6. A Lack of Wonder:
Both anecdotes portray subject matter not only as "given" but as bereft of wonder. In being expected to "accept it" as is, we are led to believe that there is little to marvel at, little to spark the feeling of surprise, little to require us to be disoriented or legitimately confused.

A humanistic orientation challenges each of these assumptions as it seeks to locate human agency in both subject matter and in educational practice. In portraying an alternative conception of the nature of teaching and learning, we need to be cautious, however, that we are not creating a straw person. While it is a helpful heuristic device to consider the consequences of negating each of the above, it is wiser to see them as bas-relief in relation to the dominant mode that they challenge. They should be seen as inspiring dialogue for the purpose of choosing how and what is to be taught and learned.

In a hasty inclination to hop along the bandwagon of seeking the negation of each of the above as the definition of "humanistic," many educators misinterpret the constructivist movement to be arguing that because all students must create meaning on their own that it makes little sense to offer lectures as a form of education. Common sense, however, tells us that we do in fact learn a great deal in ways that are not strictly "experiential." Movies, plays, and novels, for example, inspire us in many ways, despite the fact that they are being "presented." So, of course, do good lectures.

After all, such works are not merely something to be imposed; rather, they are invitations to understand another person's world, and it is the height of egocentricism to assume that we cannot learn from how it is that others see the world. It of course does not mean that we as individuals are passive when we are confronted with a play or novel or work of art or a good lecture. As a matter of fact, we have to be educated to appreciate them.[6] It also does not mean that all performances by others are appropriate as a form of education.

There are, however, times when it makes sense to accept (for a while at least) someone else's rendition of the world, to acquire a skill even when it is not clear that the skill has been born of wonder or might generate enchantments in the future. It is logically impossible to challenge or change everything at once.

Powerful dialogue is needed at every stage of education in order to decide how and when each of the above six principles (and there are of course many more that could be deduced from these anecdotes and elsewhere as well) need to be challenged, overturned, or possibly compromised by their alternatives. It is also necessary to decide when and how students are to be included in this dialogue.

The concept of human agency is one we shall be developing further in the following chapters. It will be revisited and consolidated in the final chapter. At this point, however, in order to experience the concept of humanistic mathematics education a bit more, we explore four forces or movements that exhibit a humanistic orientation: the philosophy of mathematics, the computer, the voice of progressive education and an organization (the Humanistic Mathematics Network).

The Philosophy of Mathematics: A Changing Landscape

The field of philosophy of mathematics has begun to redefine its scope. For years, its long-standing inquiry focused upon foundational interests associated with competing "isms," like logic*ism*, intuition*ism*, and formal*ism*. These various philosophical schools attempted to establish the bedrock of mathematical certainty in different ways. The schools differed in a number of ways, but they were all concerned with the issue of seeking a foundation from which everything else might flow.

These various schools mirrored for mathematics what philosophy, up through the nineteenth century, had done in its general effort to describe how and why we see the world as we do. If only we could become clear on "first principles" dealing with knowledge and truth, then we would be able to derive how and what education, for example, ought to be without having to attend in a careful way to the meanings and conflicts of educational concepts such as teaching, indoctrination, learning, and education itself.

For many reasons this program has been under attack in general philosophy and in philosophy of education since the mid-nineteenth century

Reform in School Mathematics 23

(see Scheffler, 1958). For a number of practical and theoretical reasons, the program in mathematics has also undergone some major transformations. In the early 1930s, one of the most stunning demonstrations in the foundations of mathematics itself called into question the value of pursuing some of these foundational questions. It was done in response to David Hilbert's program in mathematical formalism. He set as his agenda the task of demonstrating that mathematics as a field could not lead to inconsistencies. He also hoped to establish that we could create descriptions of mathematical systems that were complete in the sense that one would eventually be able to prove or disprove any well formed statement in mathematical systems that were at least of the order of interest and complexity of arithmetic. He hoped thus to show that if any statement could not be proven to be true or false at any particular moment in time by use of the well established axioms, then this was merely an indication of human frailty. Proving or disproving conjectures would then be a matter of commitment of time and an investment of cleverness of successive generations.

In the 1930s, Kurt Gödel, an Austrian philosopher of mathematics, demonstrated with Hilbertian rigor that both of these agendas were hopeless to achieve. Not only was the longing for a demonstration of absolute consistency found to be a pipe dream, but he also showed that in a system as ostensibly mild as the set of natural numbers, it is impossible to create a formal structure that would enable one to prove or disprove every statement belonging to it. That is, there must exist "undecidable" statements in any such system—statements that are true but unprovable as such.

How to interpret his findings and exactly what they might mean for a formal view of mathematics is a problem that philosophers of mathematics have grappled with ever since. It is clear however, that Gödel has in some sense hoisted rigor and hung it on its own petard.[7]

Rather than trying to locate the bedrock upon which all of mathematics can be based, there is an increasing inclination to seek social factors that account for a fallible view of the subject. Bloor (1991) offers his "strong programme" in arguing that sociology of knowledge can contribute not only to an understanding of how cultures and subcultures of scientific communities communicate with and influence each other, but to the logical substrata of the disciplines of mathematics and science. In a particularly revealing example, he explores how it is that what was once taken to be a totally illogical conception (the Euclidean legacy that it is impossible for the whole to be equal to any of its "smaller" parts) is revived as the central component in the definition of a concept (what it means for a set to have a countably infinite number of elements).

Lakatos (1976), in his famous *Proofs and Refutations*, demonstrates the virtual impossibility of stating relatively easy conjectures with any degree of accuracy, no less proving them. For him it is not the logical proofs of alleged theorems that advance many subfields of mathematics, but rather the eventual production of counterexamples which move the field ahead. Not only what constitutes knowledge and proof, but how they are codified and how they are passed along in the education of mathematicians are now emerging as elements of inquiry within which philosophical issues on the nature of mathematics are explored (see Kitcher, 1988; Tymoczko, 1985, 1986, 1993).

The Belgian philosopher and mathematician Van Bendegem (1993) describes how this newly emerging perspective on the nature of mathematics compares with the earlier schools of philosophy of mathematics.

> Philosophers of mathematics can be roughly divided into two types. Type I is particularly fond of questions such as: What are *the* foundations of mathematics? What are numbers? What is a set? . . . What is mathematical truth? These questions are all situated within mathematics proper. Formalists, logicists, intuitionists . . . are in this sense Type I. Type II, however, wants answers to questions such as: How is mathematics done? What is a *real* mathematical proof? . . . How is it possible that an accepted proof turns out to be wrong? Type II is still a rare species, but happily enough . . . this is changing. (p. 21)

Pedagogical derivatives of the new philosophy of mathematics have had reverberations in the emerging field of philosophy of mathematics education. Borrowing heavily from Lakatos' view of mathematics as a quasi-empirical field (neither based upon imagination alone nor exclusively upon sensory data evidence in the real world), Ernest (1991, 1994a, 1994b, 1997), Nickson & Lerman (1992), and Restivo, Van Bendegem, & Fischer (1993) are among those who see the fallible nature of mathematics as derived from the sense in which knowledge, truth and canons of proof are inextricably connected with our use of language and with social constructions of reality. Based in England, Ernest is the editor of an international newsletter that investigates a wide range of issues in the emerging field of philosophy of mathematics education.[8]

The Irony of the Computer

A popular conception of the computer is that it is a technical device that has the power to take over what heretofore required considerable human effort. It is frequently touted for its ability to outperform complicated logical analysis. The irony is that the more successful the computer has been in competing with human acts, the more we have come to appreciate the importance of human agency.

To see what is at stake here, imagine what kinds of activities we might associate with proof in mathematics. Not only do we want to (1) *attempt to prove* a conjecture (showing it to be true or false), but we also want to be able to (2) *verify* that an alleged proof in fact is legitimate. Furthermore, in order to have anything to prove in the first place, it is necessary to (3) *come up with a conjecture* that is worth proving or disproving.

Consider issue (3): *coming up with* conjectures. Fajtlowicz (1988) describes his computer program known as Graffiti, a program in graph theory that is capable of creating three to eight thousand conjectures for any collection of graphs it is handed! That is quite impressive. It is an understatement to say that Graffiti surpasses what we might expect from any human being who is handed the same information. What then is the problem? If one is handed this morass of conjectures, one has no idea what is worth pursuing and what is not. Fajtlowicz found that criteria needed to be established to determine what was "interesting," and he ended up with characteristics that of course require human judgment—ones designed to capture intuitive judgments of elegance or beauty. See Dowker (1992), Kolata (1989), and Horgan (1993) for a further discussion of the deep epistemological and educational implications of this discovery.

The same kind of problem arises with both (1) creation and (2) verification of proof. Davis and Hersh (1988) comment on a program called AUTOMATH that was devised in the 1970s for the purpose of *checking* alleged proofs. They comment, "[T]he Automath project has been virtually abandoned. There are several reasons for this. . . . Even if these translations into Automath were available in great abundance, how would one verify that they were correct, that the Automath program is itself correct, that the machine program has been correctly written, that it all has been run correctly?" (p. 68).

Back to human agency! Forward not only to the need for judgment—for a decision about what is worthwhile and significant—but for the intense realization that we cannot avoid being guided by criteria of an *aesthetic* nature that transcend logic alone when we decide what to prove, why to prove it, and whether or not it is a proof at all.

We turn now to two humanistic movements that are primarily educational in spirit. The first, progressive education, was general in nature in that it did not highlight a particular discipline. While some critics believed that the movement in fact had little truck with standard forms of knowledge, much of the writings of the person most associated with the movement, John Dewey, indicated a deep appreciation for the ways in which the disciplines were to be part of and to inform educational decisions.[9]

The second, the humanistic mathematics network, focuses primarily upon mathematics as a discipline.

The Voice of Progressive Education

Focused primarily on elementary schooling at its beginning, progressive education emerged with vigor in the first third of the twentieth-century. Often associated with—as well as misinterpreting and abusing—the American pragmatic philosophy of Dewey, it placed practice in a broad theoretical perspective that focused not only upon classroom experience but upon democratic ideals as well. Dewey's call for the application of intelligence and experience to every endeavor of human existence transformed not only the way we view education, but the ways in which we conceive of ethics, aesthetics, epistemology, and mind. He was suspicious of and reconstructed dualistic thinking of all sorts—from separating mind and body, to distinguishing between action and thinking, to seeing the religious as distinct from the secular, to distinguishing between student and subject matter.

It is helpful in gaining a further handhold on the concept of humanistic mathematics education to see how he characterized the progressive education movement. In an address to the Progressive Education Association, Dewey (1988, reprinted from 1928) reminds his audience of essential elements of the progressive mode:

- respect for individual capacities, interests and experiences;
- enough external freedom and informality at least to enable teachers to become acquainted with children as they really are;
- respect for self-initiated and self-conducted learning;
- respect for activity as the stimulus and centre of learning;
- perhaps above all, belief in social contact, communication and cooperation upon a normal human plane as an all enveloping medium. (p. 161)

Here we see the expression of a point of view that taps into essential elements of many modern programs in mathematics education—especially that of a constructivist outlook in creating knowledge and in experiencing the world, and the centrality of the social context of the classroom.[10]

In this brief summary, Dewey says little about the nature of subject matter, though he speaks in general terms about the manner in which it is to be acquired—self-initiated and through an environment of cooperation and communication. He does, however, clarify what he means by "interest and experience" and makes it clear that he does not identify himself with a starry-eyed, romantic vision of the student—one which mindlessly assumes that setting the child free to explore whatever interests him or her is the cornerstone of progressive education. Dewey (1988, reprinted from 1928) comments, "[Individuality] is something developing and to be continuously attained, not something given all at once and ready-made. . . . A child's individuality cannot be found in what [a child] does or in what he consciously likes at a given moment. It can be found only in the connected course of his actions" (pp. 164–165).

Dewey provides us with an appreciation for certain conditions that are necessary for the student to connect with the logic of subject matter. Though the teacher may need to be aware of the end product—some body of knowledge and its logical connections—the student is differently engaged. Dewey essentially warns us that the process of *coming to know* as opposed to *having achieved knowledge* requires an engaged and messy interaction with the world.

Humanistic Mathematics Network

An organization, called the Humanistic Mathematics Network, was formed in the United States at about the same time that *The Standards* were being forged. The first meeting of the organization took place in 1986 under the direction of Alvin White, a mathematician from Harvey Mudd College. He convened about a dozen university mathematicians, mathematics educators and philosophers to discuss the relationship between mathematics and the humanities. Though more heavily focused upon college/university level mathematics, the group was concerned with how the discipline of mathematics was being portrayed to students at elementary, secondary and university levels.

At that meeting, the participants created a number of tenets that defined some of its subsequent exploration. Among them were:

1. An appreciation of the role of intuition, not only in understanding, but in creating concepts that appear in their finished versions to be "merely technical."
2. An appreciation for the human dimensions that motivate discovery—competition, cooperation, the urge for holistic pictures.
3. An understanding of the value judgments implied in the growth of any discipline. Logic alone never completely accounts for *what* is investigated, *how* it is investigated and *why* it is investigated.
4. A need for teaching/learning formats that will help wean our students from a view of knowledge as certain, to be received.
5. The opportunity for students to think like a mathematician, including a chance to work on tasks of low definition, to generate new problems and to participate in controversy over mathematical issues.

6. Opportunity for faculty to do research on issues relating to teaching and to be respected for that area of research.

Since I was a participant at that session and contributed a number of the above tenets, it is not surprising that many of them cover the same territory that I described earlier as elements to be incorporated in a humanistic framework. There are some differences however. For example, though (6)— respect for faculty research on matters of education—is one that gained respect as far back as the progressive era (and has been associated in recent years with action research in grades K–12), it has been of particular concern in this group because of its composition. The focus here on research in teaching is designed largely as a political move to persuade university administrators to honor alternatives to pure mathematical research for the purpose of tenure and promotion.

Another theme that appears here that we mentioned as a slogan of the new math movement (as well as of reform associated with *The Standards*), but have not incorporated in our characterization of humanistic mathematics is (5), the opportunity for students to think like a mathematician. Though the elaboration following this description is not unreasonable, the concept is in fact more problematic than it appears at first glance. I will discuss its limitations in chapter 6.

As a consequence of the enthusiasm of this small group, a publication entitled *Humanistic Mathematics Newsletter* was begun in 1987. After seven editions, the newsletter was transformed into a journal in April of 1992. Not only has the number of subscriptions increased dramatically since its publication, but the topic of humanistic mathematics is now officially represented each year at the joint annual meeting of the Mathematics Association of America and the American Mathematical Society.[11]

In addition to the themes mentioned above (and sometimes crosshatching them), the journal is devoted to an exploration of the relationship of mathematics to the humanities in a number of philosophically interesting ways. Among the questions explored are:

- In what sense is mathematics a branch of the humanities?
- Can mathematics be reduced to one of the humanities, or vice versa?
- What is the relationship between mathematics and philosophy?
- How is mathematics expressed in the humanities (literature for example)?
- In what sense(s) is mathematics a language (and vice versa)?
- What does an awareness of the history of mathematics contribute to our understanding of the social and contextual nature of the discipline?

Consolidating and Expanding: Ambiguity of "Humanistic Mathematics Education"

What emerges in combining these various programs and movements is an intertwining of two themes, ones that are sometimes treated separately and on other occasions integrated: (1) teaching humanistic mathematics (as in teaching a view of mathematics as a meaningful human enterprise sharing many of the assumptions of other humanistic studies and experiences); and (2) teaching mathematics humanistically (as in honoring the ways in which students come to their awareness of the field).

In the case of (1), there is an appreciation for a view of mathematics as a field that is man-made and more fragile than the earlier renditions of philosophy of mathematics implied. As part of that orientation, there is an inclination to view the field as being softer on certainty, and fallible in unanticipated ways.

From the perspective of (2), there is an awareness that learning takes place through the creation and reorganization of schemes rather than in a passive mode. There is also an appreciation for the social context within which learning takes place.

Some of the pedagogical orientation is expressed in the documents affiliated with *The Standards*. As we critically analyze the uses and meanings of problems, problem solving, and applications of mathematics to the real world in the chapters that follow, however, we will be refining the concept of humanistic mathematics education beyond the bounds that have been developed so far.

In order to expand upon what we have developed so far, we shall seek clearer linkages between the transformed views of mathematics and pedagogy, (1) and (2) above. Table 1.1 provides a useful strategy for the purpose of unraveling and clarifying ambiguity and, perhaps appropriately creating it as well.

Though we ought to be cautious about viewing it all from a dichotomous point of view (and breaking each point of view further into dichotomies), as a heuristic device for thinking about the meanings of "humanistic mathematics education," it is possible to see box D as satisfying what we refer to as a radical conception of humanistic mathematics education. It combines the nature of mathematics and of pedagogy in a way that challenges the status quo. In a sound bite, box D construes learning as the construction of knowledge and mathematics as a discipline that is fallible. On the other hand, it would be possible to see box A as representing much of how we characterize the standard curriculum: learning is received knowledge, and mathematics is absolute or certain.

Table 1.1 Dichotomous depictions of the nature of mathematics and of learning.

		The Learner	
		learning as received knowledge	learning as constructed knowledge
Nature of Mathematics	math as absolute	A	C
	math as fallible	B	D

Boxes B and C represent hybrid notions of "humanistic mathematics education." In box B, we can present a view of the fallible nature of mathematics not through an active program of student discovery, but rather by passing along the wisdom of source material on the subject. In box C, we might depict students as engaged in an experiential manner for the purpose of having them come up with conclusions that are part of what we accept as belonging to the canon. This sketch is an invitation to generate further thought and to criticize the substantive issues of this book. In particular, it would be valuable to react to the way the anecdotes we used at the beginning of the section entitled "A Humanistic Perspective" depict what might be included in boxes B and C. It would be worthwhile to see how and if many of our educational concerns in the forthcoming chapters fall within this scheme.

Organization of the Book

This chapter has introduced the two main themes of the book: the concept of problem as it relates to problem posing and solving, and the notion of connection with the real world as something more vibrant than the creation of models from mathematics to the real world or vice versa. In addition, we have begun to explore the concept of humanistic mathematics as a backdrop against which the two major themes will be played out.

Part II will focus entirely on issues related to the first theme: the concept of problem. In chapter 2, we will demonstrate the sense in which problem solving is deeply embedded in the culture of mathematics, and will—by reflection on personal events—introduce its companion, problem posing. There we will begin a conversation which explores both historical and personal significance of problem posing. We will then look at problem posing—and especially problem posing associated with the What-If-Not scheme—from a number of different theoretical perspectives. We will indicate why it is so deeply built into inquiry, not just an "add-on" to problem solving. We end the chapter by introducing explicitly the relationship between the concepts of problem, solution, and situation—an idea that will receive further elaboration in chapter 4.

In chapter 3, we discuss the fallacy that has generated much of our earlier thinking about the educational potential of problems. In doing so, we will essentially be analyzing why it is that the audience in my talk to the philosophy colloquium responded to the two jeopardy game questions in the same way, despite the fact that they were in fact quite different.

Having confronted the fallacy, we are in the position to see new ways of construing problems for educational purposes. The vision, informed by the humanistic perspective introduced initially in this chapter, has some interesting "spill-over." That is, once we confront the fallacy, we need not necessarily eliminate the bond completely between problem and solution. We can take the insights we gathered about the educational agenda that flowed from focusing on problem in isolation from solution, and allow for some migration of those insights. We then can focus on different stages of relating problem to solution in connection with these new insights. We do all of this in chapter 4, and end the chapter by exploring the humanistic categories of wonder and similarity as dimensions that have the potential to influence how we see ourselves and others in relation to mathematical thought.

Part III elaborates upon the second major theme: connections of mathematics to the real world. Moving beyond the concept of model as application, we offer two other types of connection in chapter 5. In clarifying these alternative notions of "connection between mathematics and the real world," we explore further some fundamental human ways of experiencing their inextricable intertwining.

In chapter 6, we conclude the investigation of real-world connection by rejuvenating the concept of what is real. "Real" is not merely what we can "touch." It is as important to see it as what "touches us." In that context we will consolidate insights from previous chapters and will also explore alternative pedagogical strategies—using novels and a Talmudic format—a style which integrates narrative and exposition.

Part II

PROBLEM SOLVING, PROBLEM POSING, AND PROBLEMS

Chapter 2

Problem Solving and Its Emerging Companion: Philosophical and Educational Roots of Problem Posing

> Recently a teacher was overheard to announce: "When I want your questions, I'll give them to you." Much of school practice consists of giving definite, almost concrete answers. Perhaps boredom sets in as answers are given to questions that were never asked.
>
> —D. Bob Gowin from *Educating*

> I want to tell you something my brother, David, may he rest in peace, once said to me. He said it is as important to learn the important questions as it is the important answers. It is especially important to learn the questions to which there may not be good answers.
>
> —Chaim Potak, *In the Beginning*

Problem solving is deeply embedded in the culture of mathematics. In addition, it has roots that reflect a philosophical tradition which equates thinking and problem solving. In this chapter, we shall depict the centrality of problem solving in mathematics and in educational practice. The latter will be introduced by recalling some earlier innovative programs that make use of a problem solving perspective.

Though only recently acknowledged, each of these programs has within it the seeds of something more radical than a problem solving curriculum, however. We will think of these programs as a segue way to introduce the theme of the sometimes parent, sometimes sibling, sometimes offspring of problem solving: problem posing.

In moving towards problem posing, we shall inquire into the philosophical roots of one form of that activity—What-If-Not thinking. That inquiry will lead us to reconsider the role of problem posing not only in mathematical inquiry, but in its connection with a primary educational concern: that of understanding.

In chapter 4, we will further explore the more general issue of problem posing (considerably beyond the What-If-Not perspective) as an educational tool that opens pedagogical options.

Centrality of Problem Solving in Mathematics

In chapter 1 we quoted the section from *The Standards* which makes the strongest possible case not only for selecting problem solving as a primary pedagogical strategy but for viewing its absence as an indication that the subject matter itself is being compromised. Recall the comment,

"Mathematical problem solving, in its broadest sense is nearly synonymous with doing mathematics."

It is a popular view that the field progresses primarily through the act of problem solving, and to the extent that we wish to initiate others into the field, it makes sense, so the argument goes, to engage students themselves in problem solving.

What else could we do if in fact we want students to behave mathematically? The clear implication is that there may be much for students to master in the way of concepts, and there may be much that they have to acquire in the way of techniques in order to master such concepts, but unless these skills and concepts are acquired in the service of problem solving, there is little that could be defended on mathematical grounds.

Though this point of view pervades the concept of growth and of understanding in all fields, it certainly dominates mathematics. How so? Let us look first at its connection with genius in the field of mathematics. In the following section, we examine problem solving curricula that predate *The Standards* by several decades.

It is a commonplace that the concept of child prodigy is one that applies to very few fields. Though youngsters may be more talented philosophically than we generally acknowledge, we do not usually classify a ten-year-old as a philosophical or literary genius. While it is possible for a youngster to be interested in history, we do not generally identify an eight-year-old as historically gifted. In a sense these are fields that require a kind of maturity and depth that comes with age. One has to be "in the world" and experience life before being able to contribute in an original way to such fields as history, philosophy, theology, or literature. A kind of perspective must be acquired that requires an awareness of classics in the field. "What came before?" is a kind of question that is frequently acknowledged to be a critical question for those who make major contributions to the above-mentioned fields even if they eventually break with earlier traditions.

In which fields does the concept of child prodigy surface, and why is that the case? Among them are: music, mathematics, chess, and perhaps computer science in recent years. What is there about these fields that allows for such early identification? There are obvious differences among these fields, and there is also some ambiguity with regard to what sort of behavior constitutes genius in them. In music, for example, there is a difference between performing music (playing an instrument) and composing a score. It is quite possible for one to have a musical ear and to perform well even without being capable of reading music at all.

There is less ambiguity, however, in identifying mathematical genius. It is most often perceived to be closely connected with unusual feats of problem solving. Though the story of young Gauss (1777–1855), who was known as the Prince of mathematicians, is perhaps apocryphal, it illustrates an attitude of honoring genius early in one's career. His elementary school teacher, wanting relief from working with her pupils in order to perform some necessary task (like collecting milk money) supposedly asked students to find the sum of the numbers from one to one hundred. Gauss is alleged to have gotten the answer in a very short time, not by laboriously adding the numbers, as his teacher had anticipated, but rather by engaging in an act of clever problem solving—something akin to a gestalt-like shift in perspective. While everyone else saw the fact that each addend was one more than its predecessor, he saw a connection that made the problem curl up in embarrassment, as is illustrated in Figure 2.1.

Seeing that what was taken away from numbers at the end of the sequences was compensated for by a comparable increase of numbers at the beginning, he paired elements at both ends so that there was a constant sum of 101 repeated over and over again. Thus he perceived an important

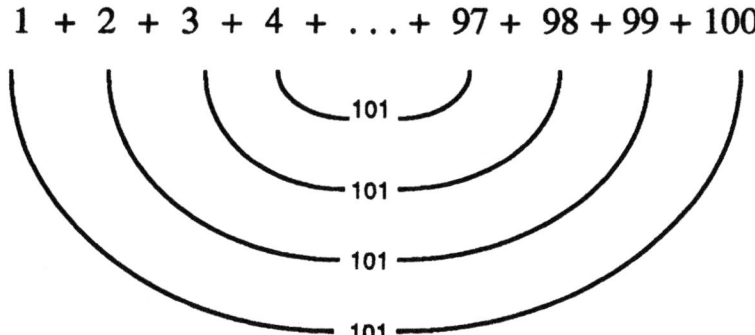

Figure 2.1 Depicting Gauss's alleged insight for adding numbers from 1 to 100.

constant among the significant variety among components of the problem—a constant that enabled him to figure out the answer (fifty pairs of 101) in his head.

It was not only that Gauss solved a problem, so the story goes, that was perceived by his classmates to be a virtual nightmare. More importantly he created a scheme which in some sense got to the (or perhaps "an") essence of the problem. He reduced something unmanageable to something manageable. He observed similarities of structure among a morass of disparate "facts." He was not distracted by intrusions that may have been true but that did not bear on the essential features of the problem.

Why does such genius emerge early in mathematics and not so in what we normally conceive of as the humanities? Though we should be cautious about overgeneralizing from this instance of talented behavior, the situation does have important components that may contribute to an understanding of the centrality of problem solving in mathematics.

For one thing the "machinery" needed for Gauss to exhibit his brilliant insight is relatively meager. That is, it was not necessary for him to know a great deal in order to put his ideas together in a clever and novel way. There was very little in the way of technical and formal understandings that were prerequisite to solving the problem. Though it was clever indeed, it was an unnoticed insight that was accessible to almost anyone who had experienced the adding of numbers. Though Gauss had surely added numbers before, as had all of his classmates, he most likely had thought of numbers and of operations on them differently. He most likely had viewed calculation problems not merely as tasks to be performed but as an invitation to see their dynamic relationships.

A second feature of the mathematical talent Gauss demonstrated is that the context within which he perceived the problem was essentially local and self-contained. It was local in the sense that it was not necessary to know where the problem came from, why it was given, why it was worth thinking about, and what else flowed from it. It was self-contained in that it was perceived as a kind of experience that was not closely related to other experiences.

Mathematical genius is an intriguing concept in that it need not necessarily be focused so sharply on underlying structure or fundamental properties of a system as was the case with Gauss's discovery. In fact, it can flourish in an environment of particularity and the uniqueness of context as well. There is a wonderful story told by David Hardy (1877–1947), a leading British mathematician about Ramanujan, an Indian mathematician who made groundbreaking but untutored observations, frequently using unconventional symbolism. Newman (1956) quotes Hardy as saying about Ramanujan, "I remember once going to see him when he was lying ill at Putney. I had ridden in taxi-cab number 1729, and remarked that the number seemed to me a rather dull one, and that I hoped it was not an unfavorable omen. 'No,' he replied, 'it is a very interesting number; it is the smallest number expressible as a sum of two cubes in two different ways'" (p. 375).

It is hard to know what was going on in Ramanujan's thinking that enabled him to come up with this observation. It is surely possible that he solved no problem at all in arriving at the observation that 1729 had a very unusual property. He may have "merely" recalled an observation he had made in his many previous calculations. It is possible that a lot of his thinking is of a sort that is inaccessible and virtually impossible for outsiders to discover—much as the mythology goes with regard to *idiot savants*. It is just as likely, however, that he was so immersed in thinking about numbers and about their relationships in an untutored way that he had created an enormous number of connections that enabled him to capture any particular number in his messy and inefficient but aesthetically appealing web.

Regardless of the manner in which the two concepts of genius appear, they do share features of isolation and simplicity, and in addition they do not require a sort of maturity (and frequently personal angst) that we associate with fields such as literature, philosophy, theology, and history. It is of course a far cry from the depiction of genius to selecting those qualities as the essential definition of a field. It is also a far cry to forge an educational agenda from such depiction.

In an important sense, there is nothing God-given about identifying mathematical genius with either the Gauss or the Ramanujan model of problem solving, and much of this text is an effort to reconstruct that dominant point of view. Kincheloe, Steinberg, & Tippins (1992) argue that the concept of genius is one that is socially constructed and not inherent in the structure of a field. This is an issue we elaborate upon in chapter 6 in discussing the concept of talent derived from the empirical findings of Krutetskii (1976).

If we do not necessarily adopt the traditional earmarks of genius directly as a defining goal of curriculum, we perhaps find it appealing to educate "ordinary" youngsters in a manner that exemplifies its qualities at a lower level of sophistication. To the extent that this is the case, it is that much more critical that we seek alternative and more humanistic modes of mathematical thinking and behavior that would redefine what is talented behavior.

We already presented a view not of genius, but of mathematical progress in chapter 1 that puts reins on problem solving. Lakatos' argument regarding the power of counterexamples does suggest the need to examine the perception that it is problem solving (as depicted in this section) that unfolds truths in some systematic way that illuminates underlying schemes. In chapter 4, we will suggest a notion of progress that calls into question the centrality of a narrow conception of problem solving ("Problem / Situation Reverberation: A Return to Progress in Mathematics"). Nevertheless, this narrow notion is a powerful one, and the culture of mathematics seems to operate as if it is the driving force. In the next two subsections, we will demonstrate its centrality in mathematics education for a longer period of time than we have been led to believe. We follow these subsections with a form of deconstruction that reveals that another component has been present but suppressed as an explicit ally of problem solving, especially in educational contexts: that of problem posing ("Problem Posing: An Implicit and Emerging Voice").

Pedagogical Models Predating the Reform Movement

If problem solving is touted as a central feature of the culture of mathematics, it has surely been deeply implicated in mathematics education long before the voice of *The Standards*. As a matter of fact, celebrating problem solving as if it were a newly discovered educational commodity is strangely reminiscent of the story of the emperor's new clothes. For the purpose of reminding us of its long-standing tradition, but also as a precursor to an interesting twist that will become explicit in the next major section, I shall describe briefly two innovative programs, one from the secondary school and one from the college scene. Both predated the present movement by over half a century.

A Secondary-School Example
One of the most interesting experimental teaching designs that involved problem solving in relation to proof took place in the mid 1930s. Concerned with the realization that students were being taught geometry as their introduction to formal deduction, but painfully aware of the fact that most of them viewed the activity as an arbitrary collection of statements and reasons, Harold Fawcett reorganized his teaching. Instead of imposing a set of Euclidean theorems as a finished product to be learned in some predefined sequence, he began by asking his students to tell him everything they could about points, lines, and planes based upon their

intuitions about how the world behaved. Having amassed such "data" as a starting point, he then had each student figure out which statements seemed more obvious than others, which might be hard to accept, and which might in fact be false. Fawcett (1938) comments, "Since the student is to have the opportunity to reason about the subject matter of geometry in his own way, no definite sequence of theorems can be arranged in advance" (p. 24). The students then began an adventure in which they defined what they would explore, and they did so in such a way that they became aware of the role of deduction not as an arbitrary act, but rather as one that could affect their own beliefs about statements that they had generated.

As part of what was an essentially constructivist environment par excellence, no text was used in the course. Rather each pupil wrote his or her own text and was able to "express his own individuality in organization, in arrangement, in clarity of presentation and in the kind and number of implications established" (p. 62). In focusing on method rather than on substance per se, Fawcett intuitively appreciated a distinction—the critical difference between viewing mathematics as a body of knowledge and as a way of thinking.

Fawcett's intentions were more ambitious however than having students appreciate the nature of deductive thought in mathematics alone. Fawcett (1938) comments:

> While teachers of mathematics say they want the young people in our secondary schools to understand the nature of proof, that should not be and probably is not their total concern. What these teachers really want is not only that these young people should understand the nature of proof but that their way of life should show that they understand it. Of what value is it for a pupil to understand thoroughly what a proof means if it does not clarify his thinking and make him more "critical of new ideas presented"? The real value of this sort of training to any pupil is determined by its effect on his (sic) behavior, and for purposes of this study we shall assume that if he clearly understands these aspects of the nature of proof his behavior will be marked by the following characteristics:
>
> 1. He will select the significant words and phrases in any statement that is important to him and ask that they be carefully defined.
> 2. He will require evidence in support of any conclusion he is pressed to accept.
> 3. He will analyze that evidence and distinguish fact from assumption.
> 4. He will recognize stated and unstated assumptions essential to the conclusion. He will evaluate these assumptions, accepting some and rejecting others. (p. 11)

Thus Fawcett's agenda extends to an ability to engage in some form of critical thinking as well.

From the College Scene

An equally innovative program whose major focus was on problem solving appeared on the college scene around the 1930s as well. R. L. Moore, teaching at the University of Texas, organized his classes completely around students' attempts to make sense out of problems that he introduced throughout the course. Some of the propositions he presented were in fact false; some were true and well known; some were problems that had been heretofore unsolved. Students never knew from which of these categories their problems came.

The problems were not presented in the spirit of homework to be done in order to reinforce some ideas that were independently being developed. The problems were the substance of the course. Furthermore, the student's role was clearly circumscribed in this regime. That is, anyone who at any point felt she or he had a solution to one of the problems would be invited to present it to the class. Moore's role as well as that of the other students was to act solely as a critic of the alleged solution. More precisely, a student could point out that there was something wrong with it, and perhaps even indicate what the specific difficulty was, but was prohibited from trying to rectify an error that someone else had committed. R. L. Moore claims that to do so would rob students of the satisfaction of solving a problem on their own. Students were even encouraged to leave the room when a solution to a problem on which they were working was being presented by another student in the class.

This model not only established a sense of independence rather than cooperation in the classroom itself, but created an atmosphere in which all outside authority was suppressed. Students were not only discouraged from receiving constructive help in class, but they were forbidden from communicating with each other outside of class and were not permitted to consult texts or other instructors.

What might be viewed as highly suspect (if not downright immoral and debilitating) in the light of present-day prizing of cooperative learning, was a life-enhancing source for an unusually large number of students. In fact, R. L. Moore single-handedly turned out this century's leading set-theoretic topologists. Moïse (1965), a student of his, describes the situation as follows:

> Most of the time . . . a student of Moore doesn't even know whether the problems that he is working on are research problems. Eventually, they turn out to be, and the result is a thesis. Thus, at the time when I wrote my first research paper, I had never read one. And I had never discussed topology with fellow-students outside of class. If this scheme of teaching seems misguided, we should remem-

ber that the list of Moore's students includes the names of R. L. Wilder, G. T. Whyburn, R. H. Bing, *and many other* research mathematicians of distinguished achievements. . . . And Moore's record as a teacher is even more impressive in light of the circumstances under which he has taught. There are some universities in the United States where a professor can take for granted that the best and most ambitious graduate students in the country will arrive, every year, already well trained and already fully committed as mathematicians. The University of Texas is not one of these, and so Moore has had the task of recruitment as well as the task of teaching. (p. 408, emphasis in original).

Problem Posing: An Implicit and Emerging Voice

We have chosen two fascinating examples of the functioning of problem solving in the culture of mathematics. They serve the purpose not only of suggesting that some interesting radical teaching models based upon problem solving have been around for a while, but they also remind us of a "near relative" of problem solving that has been suppressed and only recently acknowledged: problem posing. We begin our analysis of this missing ingredient with a personal anecdote that occurred in the mid 1960s when the new math was still flowering.

> I was team-teaching a course on problem solving to beginning teachers together with my colleague Marion Walter. In an early stage of the course, we were interested in having the students investigate properties involving triples of numbers known as Pythagorean Triples, the most famous of which is 3,4,5. These sets of triples are whole numbers with a very special property. From a geometric point of view, when a triangle is drawn whose three sides have the lengths of an associated triplet (as is the case with 3,4,5), it turns out that the triangle must have a 90-degree angle (a right angle). Algebraically, such triples (x,y,z) are whole numbers that satisfy the equation:
>
> $x^2 + y^2 = z^2$
> e.g., ($3^2 + 4^2 = 5^2$)
>
> In an effort to involve students in generating such triples in a spirit of inquiry, we asked them the following question:
>
> $x^2 + y^2 = z^2$ What are some answers?
>
> We received a number of responses to our question. Among them were:
>
> 3, 4, 5;
> 5, 12, 13;
> 8, 15, 17;
> 6, 8, 10.

Eventually, a number of students, thinking they would pull our leg came up with answers like:

0, 0, 0,
1, 1, 2,
−3, 4, 5.

It was only after class ended, and we began to discuss what had occurred that we realized that the students had not pulled our leg long enough! The difficulty was not that there might be even other extreme cases that they had not suggested (like imaginary numbers) but a much more interesting issue was at stake.

The phenomenal thing about that experience was that they were supposedly solving a problem when in fact no problem had been posed!

$x^2 + y^2 = z^2$ is itself neither a problem nor a question capable of being answered. In fact a most appropriate response to the request for answers would have been, "What's the question or the problem?"

(adapted and modified from Brown & Walter, 1983, pp. 9–10; and Brown & Walter, 1988, pp. 123–124).

Nothing about the behavior of the students is surprising. Such short-circuiting of dialogue is part of what is involved in developing a common language among people in a particular field of inquiry. By informal convention $x^2 + y^2 = z^2$ is frequently a signal that people who share that language are searching for or dealing with Pythagorean triples. This is a good argument in defense of the students who had produced the original list of Pythagorean Triples.

Such short circuiting, however, is problematic for a number of reasons when carried out with people who are first becoming familiar with a body of knowledge, people who do not yet understand the boundaries of a field. Furthermore, a regular diet of such short circuiting has the effect of curtailing a great deal of inquiry—inquiry that is sometimes experienced more intensely and intelligently among neophytes than those who are deeply entrenched in a field. Even when one's teaching goal is to lead students up well traveled paths, it makes a great deal of sense to encourage them to explore side paths that may sometimes appear and even be irrelevant as a means of achieving such predetermined goals.

I see this experience as one that precipitated an awareness that has had an enormous impact on the way I think about problem posing and problem solving. Problem solving is so deeply implicated in the culture of mathematics and mathematics education that students are inclined to view what they are expected to do as a form of problem solving even when no problem has been posed! So began a journey (in the mid-1960s) of many years in which I explored a number of pedagogical, personal, and logical issues that linked the two activities.[1]

Problem Solving and Its Emerging Companion

The language of problem posing has in fact been included in the agenda of the recent national documents that focus on problem solving. While *The Standards* do suggest the inclusion of problem posing, the recommendations, though important, are minimally analyzed and not adequately connected with other important educational/mathematical themes. There are, for example, suggestions in *The Standards* that students be given some information, and then asked to pose problems as in the following commentary from the National Council of Teachers of Mathematics' (1989) *Curriculum and Evaluation Standards for School Mathematics*:

> Students in grades 9–12 should also have some experience recognizing and formulating their own problems, an activity that is at the heart of doing mathematics. For example, exploration of the perimeters of various rectangles with area 24 cm² by means of models or drawings, with data as recorded in [the table below], could lead to student recognition and formulation of such problems as the following: Is there a rectangle of minimum perimeter with the specified area? What are its dimensions?

Rectangle Data

Area	Length	Width	Perimeter
24 cm²	1 cm	24 cm	50 cm
24 cm²	2 cm	12 cm	28 cm
24 cm²	3 cm	8 cm	22 cm
24 cm²	4 cm	6 cm	20 cm
24 cm²	6 cm	4 cm	20 cm
24 cm²	8 cm	3 cm	22 cm

Much of the rationale for such inclusion in documents such as *The Standards*, however is consistent with the early stages of my own awareness of the connection. There is a great deal more to be thought about from an educational point of view than appears in these early commitments. I shall speak first about some of these initial connections and will subsequently expose what I consider to be deeper ones that have taken almost a quarter of a century to enter my awareness. In doing so, I will be further expanding upon the foundation that was developed in my earlier writing on the subject.

Problem Posing: Initial Explorations of Its Significance

Why is problem posing important and how might it relate to problem solving in mathematics education? As a start, it is worth acknowledging two important issues about problem solving. First of all, no problem is solved without first being posed. This does not mean that all problems are well posed or that they do not undergo some metamorphosis in their development, but rather that it makes no sense to speak about *solving* something that has not been *posed* as a problem. Sometimes the problem is implicit, as in fact it was in the case of the search for Pythagorean triples, but something has to be there to receive a solution. Furthermore, it is important to appreciate that posing problems is not a simple task. From the point of view of the history of a field as well as from the point of view of the individual's development, it frequently takes a long time before problems are posed in such a way that their formulation is capable of advancing inquiry (if that is the goal).

Choosing a more popular, and perhaps timely, problem, imagine the differences among the following three formulations:

1. What kind of protection should I devise so that I minimize the chances that someone will break into my house?
2. Do I want to live in a community in which theft and violence are rampant?
3. What obligation do I have to myself and others to try to modify the violent climate in my neighborhood?

It is interesting to speculate on whether or not we should consider these to be different formulations of the same problem or merely different problems in need of solution. However, the way in which we ask the question has enormous consequences for how we lead our lives. We also should be aware of the fact that we do not always know before we begin to inquire which formulation of the problem is one that "really" drives us. Much of therapy is about discovering that what we thought was a good formulation of a problem in fact was masking what really was bothering us.

As in the case of personal and interpersonal reflection and introspection, the history of any field reveals a similar need to be sensitive to the way in which problems are posed. Perhaps the most stunning example of the resistance of a solution to a problem by virtue of the way in which it was posed can be found in the history of the parallel postulate.

Among the postulates Euclid stipulated as essential elements over two thousand years ago in the foundation of plane geometry are the following (derived from Heath, 1956):[2]

> Postulate 1: A straight line can be drawn from any point to any other point. (p. 195)
> Postulate 2: A straight line can be extended indefinitely in both directions. (p. 196)

For years these postulates were considered to be "self-evident." Another postulate of Euclid's, however, was considered to be more complicated than the others so that mathematicians believed that it could be derived from the others. That is the fifth postulate of Euclid, which is:

> Postulate 5: If a straight line falling on two straight lines makes the interior angles on the same side less than two right angles, then the two straight lines, if produced indefinitely, meet on that side on which are the angles less than the two right angles. (p. 202)

The postulate is depicted in Figure 2.2.
In Figure 2.2, line 1 is "a straight line" and lines 2 and 3 are the "two straight lines" mentioned in postulate 5. Angles 1 and 2 are the interior angles and P is where the lines meet.
This postulate is equivalent to a more popular version of the postulate, which says:

> Parallel Postulate: Through a given point not on a line, there is exactly one line parallel to the given line.[3]

The long history of non-Euclidean geometry was essentially an attempt to show that the parallel postulate was less primitive than the others. Efforts were made to derive that postulate from the earlier ones of Euclid. One of the most interesting digressions in the long history of the field is due to the work of an Italian geometer, Gerolamo Saccheri (1667–1733). He published a book entitled, *Euclides Ab Omne Naevo Vindicatus,* in which he attempted to "vindicate Euclid of every blemish" by finally coming up with a proof of the parallel postulate.

Moïse (1963) describes the high humor associated with Saccheri's argument. First of all, there turned out to be a major flaw in his logic

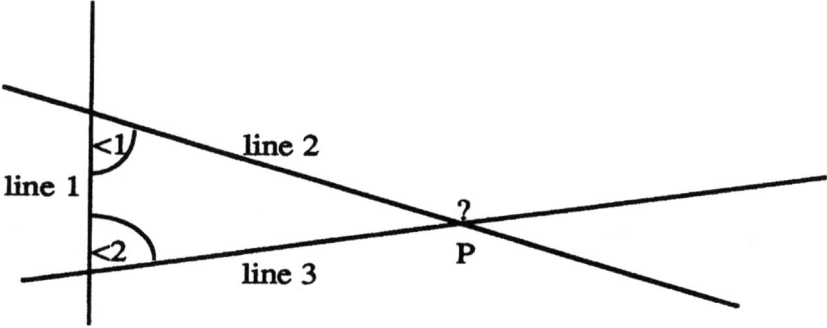

Figure 2.2 Postulate 5 of Euclid's Elements.

(having assumed without realizing it that he was making use of properties of straight lines that were not warranted by the postulates alone), and in fact he had not established the postulate as a theorem at all. Second Moïse argues, "The final irony is that if Saccheri's enterprise had succeeded in the way he thought it had, no modern mathematician would have regarded his book as a vindication of Euclid. From a modern viewpoint, a proof of the parallel postulate would merely show that the postulate was redundant; and redundancy is not thought of as a virtue. . . ." (p. 131).

In the mid-nineteenth century, the problem of the parallel postulate was finally solved. It was solved in a way that truly vindicated Euclid according to what were then modern canons of mathematical logic. It was shown that the parallel postulate was independent of the other postulates. That means that it was proven that the postulate *could not* be derived from the others.[4] Thus began the history of systems in which forms of the following assumptions were stipulated in different renditions of geometry:

1. Given a line with a point not on the line, there are many lines parallel to the given line.
2. Given a line with a point not on the line, there are no lines parallel to a given line.

Why, however, was this problem of establishing the role of the parallel postulate so resistant for centuries? The answer in large part is that the problem being posed for investigation was flawed. That is, the question that was being investigated was, "How can we prove the parallel postulate from the other ones?"

What precisely is wrong with the above formulation? Its flaw (at least according to many modern child-rearing theorists) is analogous to asking "How can we appropriately punish our teen-age children so that they will grow up to be good citizens?"

If in fact punishment is viewed to be an inappropriate form of child-rearing, then no amount of effort to seek an answer that makes a pro-abuse assumption will be productive. Similarly, one has to take a more cautious view of the parallel postulate question in order to achieve progress. What has to be tossed away is the little word "how" that begins the question in both the child-rearing and the geometric cases. Once formulated without the "how," there is a chance the problem can be answered (in the negative). In other words, the question that eventually led Bolyai, Lobachevsky, and others to crack the problem of the parallel postulate in the middle of the nineteenth century was, "*Can* (not 'How can') we prove the parallel postulate from the other ones?"

Now of course, it would have been possible to ask the question with the "how" inserted and still make headway by showing that the question cannot be answered as asked. The asking of the "flawed" question in the way it was asked, however, was not a frivolous matter. It in fact was an indication of a deeply held belief that the postulate was in fact provable.

Why is it that we hold onto our assumptions (like the one that the parallel postulate is provable) and cannot imagine alternatives? What does it finally take to admit that the way in which a question is being asked is the culprit? There are many different answers to these questions. Kuhn (1962) argues that in scientific enterprises, there exist different paradigms that people of different persuasions adopt (on the basis of training, for example) and that these paradigms are in some important sense incommensurable. He argues that it may be less a question of experimentation and debate among those adopting different paradigms that results in finally adopting one theory over another. Rather he speaks of "scales falling off the eyes" among the competition and of new generations being trained and older ones dying off.

There are other sorts of explanations for why it is that we may not be able to adopt new paradigms, no less "see" evidence when it supposedly is staring us in the face. Many proponents of a postmodern tradition generalize from Kuhn's arguments about the nature of scientific thinking and conclude that we may not be able to reduce differences in point of view to categories that can be decided upon rational grounds. Others argue that Kuhn has confused some vital categories of scientific inquiry. Many philosophers of science claim, for example, that although there may not be canons that account for *discovery* of new ideas, the issue of *verification* is another matter. (The inclination to see the two perspectives as unrelated is one that has been challenged in recent years by a feminist epistemology of science).[5]

In the next few sections, we will elaborate on philosophical and educational issues that focus on a quite special form of problem posing: What-If-Not thinking. We will show how it is deeply implicated in our efforts as individuals and as a culture to understand new ideas. Furthermore, it has the potential to confront some of the crusty problems that Kuhn has argued are tantamount to "scales falling off the eyes."

What-If-Not and Problem Posing: Philosophical Issues

A problem posing strategy that appears on the surface to be relatively simple, yet which gets at the deepest of human cognitions and emotions involves asking "what-if-not"? [6] It is an activity that is counterfactual and it is one that may take centuries to even ask. We have already discussed an example of such intransigence in the case of the parallel postulate. In other words, it took thousands of years for mathematicians to have the vision and courage to ask what the consequences would be if the parallel postulate were not provable from the other postulates despite the fact that it looked on the surface to be considerably more complicated than the others and thus perhaps derivable from the others.

There are numerous practical and theoretical issues to investigate with regard to the What-If-Not strategy. Kenneth Burke (1968), in exploring the widely held belief that what distinguishes humans from other species is their symbol-making ability, puts a fine point on that perspective when he asserts that the most significant part of the symbolic act is the invention of the negative. To behave in a way that accords with commandments "not to do certain things" is quite different, he points out, from being able to appreciate the actual function of negative thought—an act that is propositional and connected intimately with an appreciation for the role of language. He says, "When first working on the negative, I thought of looking through the documents on the training of Helen Keller and Laura Bridgeman, whose physical privations made it so difficult to teach them language. And in both cases, the record showed that the hortatory negative was taught first, and it was later applied for use as propositional negative, without explicit recognition of the change in application." (p. 10)

With regard to the functioning of negative thought, he makes a powerful observation that has many implications for the use of the What-If-Not strategy. That is, the concept of negativity involves unfulfilled expectations. He comments,

> If I am expecting a certain situation, and a different situation occurs, I can say that the expected situation did *not* occur. But so far as the actual state of affairs is concerned, some situation positively prevails, and that's that. . . . I can ask, "Does the thermometer read 54?" And if it registers anything in the world but 54, your proper answer can be, "It is not 54." Yet there is no such thing as its simply *not* being 54; it is 53, or 55, or whatever. (p. 10)

The distinction in meaning between the following two propositions is surely significant:

(a) The thermometer tells us that the temperature is 54 degrees.
(b) The thermometer tells us that the temperature is not 54 degrees.

While I agree with Burke's observation about the power of negation, I disagree with his assertion that there is automatically a significant difference in the *amount of information* conveyed by each. Thus, we cannot conclude, as Burke implies that (b) ipso facto is less precise than (a). First of all, by virtue of their propositional status, we are in a position to seek either truth or falsity in each case. Second, we need to know more before we pass judgment on the relative precision of the two statements.

There surely may be cases, as Burke suggests, that once we observe that the temperature is 54 degrees and then associate that observation with proposition (a), we might wonder what would be the consequences of negating proposition (a). We might even have someone tell us (and believe it to be true) that proposition (b) is the case. What, however, would we be inclined to do if we believed such information?

We might, as Burke suggests, feel that the information is inadequate and that we need to conduct inquiry of various sorts that could hone in on a more precise statement, such as:

(c) The thermometer tells us that the temperature is precisely 55 degrees.
(d) The thermometer tells us that the temperature is greater than 54 degrees.

We are not, however, automatically drawn to investigate a (c) or (d) kind of propositions if we believe that (b) is the case. Burke neglects to appreciate that the kind of inquiry we conduct depends largely upon our intention. If, for example, we know that the heating system in our house gets stuck and no longer works on its timed program if the temperature at any point registers 54 degrees, then we might find it quite adequate to be told that proposition (b) is the case. The information we receive would be sufficient for our purposes (e.g., we realize we have to call a repair person), and we would not feel the need to conduct further inquiry.

There are many purposes, however, for which the information of proposition (b) would generate new inquiry—inquiry that would encourage us to investigate (d) or perhaps even something that we would consider more precise, like (c) or a variation of it. Furthermore, there is no easily accessible heuristic that would tell us what we should focus on in attempting to gain additional information with regard to a negated proposition. Thus, there is no way for us to bypass questions of *purpose* in our conducting inquiry and consequently in deciding which of several competing propositions may be true.

Douglas Hofstadter (1985) is similarly intrigued by the act of being able to ask a what-if-not question. He focuses not so much on its linguistic and propositional quality (though it is implied by his observation), but upon the ability to even imagine what is not there. He believes that such activity is not only a distinguishing feature of humankind, as Burke claims, but that it is at the crux of all creative behavior.

He is intrigued by the quote from Shaw's *Back to Methuselah* (a quote frequently attributed to Robert Kennedy):

> You see things; and you say "Why?"
> But I dream things that never were; and I say "Why not?"

Picking up on Shaw's quote, Hofstadter comments, " 'To dream things that never were'—this is not just a poetic phrase, but a truth about human nature. Even the dullest of us is endowed with this strange ability to come up with counterfactual worlds and to dream. But why do we have this ability—in fact, this proclivity? What sense does it make? And—how can one 'see' what is visibly *not there?*" (p. 232). He no sooner asks the question than he moves, with some degree of discomfort, to the hypothesis that imagining *what is not there* is the crux of creativity. His realization of the radical nature of such a move is expressed below:

> On the face of it, this thesis is crazy. How can it possibly be true? Aren't variations simply derivative notions, never truly original creations? Isn't the notion of a 4X4X4 cube simply a result or "twiddling a knob" on the concept of Rubik's-Cubicity? You merely twist the knob from its "factory setting" of 3 to the new setting of 4, and presto—you've got it! An inner voice protests: "That's just too easy. That's certainly not where Rubik's Cube, the *Rite of Spring*, relativity or *Romeo and Juliet* came from, is it? Isn't there a 'magic spark' that leaps across a gap when a Rubik or a Stravinsky or an Einstein or a Shakespeare comes up with a great idea, something that is patently lacking when an Eve Rybody merely twiddles a knob on an already-existing notion?" (p. 233)

The powerful insight that he comes to, however, is one that is implied in our earlier discussion of the parallel postulate. Merely "tweaking" something to change it into something else requires an act that is much deeper than that of "tweaking" alone. It requires *noticing* that there is something to tweak. He comments, "Making variations is not just twiddling a knob before you; part of the act is to manufacture the knob yourself. Where does a knob come from? The question amounts to asking: 'How do you see a *variable* where there *is* actually a *constant*?' More specifically: *What might vary*, and *how* might it vary?" (p. 250, emphasis in original).

Part of the educational agenda in connection with problem posing is to teach people to *notice* what they had not noticed before and part of what is needed in order to encourage such noticing is a full-blown use of What-If-Not thinking, not just in connection with problem solving but as an activity that has a value of its own.[7]

Dewey, in focusing not only on creativity but on the nature of thought—especially in the field of philosophy—also sees a What-If-Not perspective as an essential ingredient of inquiry, although he does not use that language. He points out that classical philosophy, which viewed this world as an imperfect version of something grander, strove to locate the essence of reality. As in Plato's image of the cave, the world we all experience provides only a glimpse of the perfect world whose images might be caught through reflected sunlight in an otherwise dark cave. The goal of classical philosophy then was to use imperfect examples from this world as a heuristic to derive abstract essences—essences of justice, beauty, and the like.

Our world that is filled with change from one minute to the next and is essentially unstable was not according to earlier philosophers the world that one hoped to comprehend. Dewey (1920), picking up on an inclination of poets and connecting that point of view with the growth of philosophy comments,

> What is the chief source of the complaint of poet and moralist with the goods, the values and satisfactions of experience? Rarely is the complaint that such things do not exist; it is that although existing they are momentary, transient, fleeting. They do not stay; at worst they come only to annoy and tease with their hurried and disappearing taste of what might be; at best they come only to inspire and instruct with a passing hint of truer reality. This commonplace of the poet and moralist as to the impermanence not only of sensuous enjoyment, but of fame and civic achievements was profoundly reflected upon by philosophers, especially by Plato and Aristotle. The results of their thinking have been wrought into the very fabric of western ideas. Time, change, movement are signs that what the Greeks called Non-Being somehow infect true Being. The phraseology is now strange, but many a modern who ridicules the conception of Non-Being repeats the same thought under the name of the Finite or Imperfect. (pp. 106–107)

Thus the study of philosophy was understood to be a contemplative act, one that focused not upon the here and now but upon the ideal world. It was by contemplation of the ideal that one achieved an understanding of the nature of humankind. Having a deep appreciation for the influence of modern science, however, Dewey points in another direction for gaining a better understanding of the world. He says,

> Let us turn . . . from this conception of the measure of true knowledge and the nature of true philosophy to the existing practice of knowledge. Nowadays if a man, say a physicist or chemist wants to know something, the last thing he does is merely to contemplate. He does not look . . . upon the object expecting that thereby he will detect its fixed and characteristic form. He does not expect any amount of such aloof scrutiny to reveal to him any secrets. He proceeds to *do* something, to bring some energy to bear upon the substance to see how it reacts; he places it under unusual conditions in order to induce some change. While the astronomer cannot change the remote stars, even he no longer merely gazes. If he cannot change the stars themselves, he can at least by lens and prism change their light as it reaches the earth; he can lay traps for discovering changes which would otherwise escape notice. Instead of taking an antagonistic attitude toward change and denying it to the stars because of their divinity and perfection, he is on constant and alert watch to find some change through which he can form an inference as to the formation of stars and systems of stars. Change in short is no longer looked upon as a fall from grace, as a lapse from reality or a sign of imperfection of Being. Modern science no longer tries to find some fixed form or essence behind each process of change. Rather, the experimental method tries to break down apparent fixities and to induce changes. (pp. 112–113)

Putting his finger precisely on one factor that motivates the centrality of an educational program which focuses upon a What-If-Not point of view, he comments,

> . . . In short, the thing which is to be accepted and paid heed to is not what is originally given but that which emerges after the thing has been set under a great variety of circumstances in order to see how it behaves. Now this marks a much more general change in the human attitude than perhaps appears at first sight. It signifies nothing less than that the world or any part of it as it presents itself at a given time is accepted or acquiesced in only as *material* for change. (p. 114, emphasis in original)

Picking up on the concept of what it means for something to be scientific, as opposed to being "pseudoscientific," Karl Popper (1959) also locates a What-If-Not point of view as central. He asks the pointed question: What is it that makes a theory scientific? It is not only defective science but the unscientific nature of religion and poetry that he tries to identify.

Many people would argue that what lends credibility to our belief in the legitimacy of a scientific theory would be our ability to find positive instances of the theory. Popper, however, points out that in many fields of inquiry, we have set up the way in which we inquire so that it would be impossible to find a negative instance. If, for example, one adopts the point of view that everything happens for the best, then what might we use as evidence that such a theory is defective? The answer is nothing. If disaster occurs, for example, then that does not disprove the theory. It is rather an example of the overarching belief either that evidence of its good consequences may someday be forthcoming, or that it may already exist, but that we may never be in a position to know or see it.

Using the rather simple test of negativity, Popper (1959) comes up with an analysis of what it means for something to be scientific. His intention is to criticize the positivistic school of thought that equates meaningfulness with the claim that all scientific propositions must have the potential to be shown to be true or false —the logical potential of both truth and falsity given equal weight. He comments,

> [T]he positivistic dogma of meaning . . . is equivalent to the requirement that all the statements of empirical science (or all "meaningful" statements) must be capable of being finally decided, with respect to their truth *and* falsity; we shall say that they must be *"conclusively decidable."* This means that their form must be such that *to verify them and to falsify them* must both be logically possible. . . . [Thus] Waismann says . . . clearly: "If there is no possible way to *determine whether a statement is true* then that statement has no meaning whatsoever. For the meaning of a statement is the method of its verification." (p. 40, emphasis in original)

It is not so much that we must be able to find instances that are negative in that they would refute the statement, but rather that we must be able to imagine that such evidence *could* exist. That is, the statement (or the theory if one wants to speak in broader terms) must have the *potential* to be falsifiable. He comments further,

> [N]ot the *verifiability* but the *falsifiability* of a system is to be taken as a criterion of demarcation. In other words: I shall not require of a scientific system that it shall be capable of being singled out, once and for all, in a positive sense; but I shall require that its logical form shall be such that it can be singled out, by means of empirical tests, in a negative sense: *it must be possible for an empirical scientific system to be refuted by experience.* (p. 41)

It is the inability of grand theories like much of Freudian psychology and astrology to be falsified that leads Popper to conclude that although such theories may serve important human functions (for example, they might inspire us to act in certain positive ways), they lack the most basic element of scientific thinking.

What-If-Not: Educational Matters

What-If-Not thinking has been selected by a number of different philosophical perspectives as a way of thought that is (1) peculiarly human, (2) that characterizes the essence of creative thought and, (3) that is at the center of our effort to determine the meaningfulness if not the scientific nature of statements and theories about the world.

There is a sense, however, in which such thinking is even more intimately connected with an educational agenda. That is, if one asks what it is we want students to be able to do with the knowledge and experiences we offer them in strictly educational settings, most teachers would claim that what they want is for students to *understand* what they are learning. Regardless of how much they might argue for the necessity of "knowing the facts" as a precondition for being able to understand what they are learning, most educators would claim that the achievement of understanding is the ultimate goal. They would claim, for example, that they do not want them to merely "parrot" what they are taught.

On Understanding: An Enticement

In order to gain some handhold on the nature of the concept of understanding in relation to What-If-Not thinking, we turn to a rather innocent sounding observation by Henri Poincaré (1961), a leading French mathematician of the early twentieth century. He observes and then queries as follows:

> If mathematics invokes only the rules of logic such as are accepted by all normal minds; if its evidence is based on principles common to all men, and that none could deny without being mad, how does it come about that so many persons are here refractory?
>
> That not everyone can invent is nowise mysterious. That not everyone can retain a demonstration once learned may also pass. But that not everyone can understand mathematical reasoning when explained appears very surprising when we think of it. And yet those who can follow this reasoning only with difficulty are in the majority: that is undeniable, and will surely not be gainsaid by the experience of secondary school teachers. (p. 33)

Surely he is correct in asserting that secondary school teachers would not deny that merely following mathematical reasoning once presented is a talent that is more difficult to acquire than one might imagine. However, there is something radically wrong with the point of view that Poincaré expresses. What is it?

We might wish to take issue with his original observation that mathematics is essentially about following rules of logic. Even disregarding sexist sensitivity, we might confront his assertion that its evidence is based on principles common to all men. There is, however, something more devastating that he accepts which is in need of deconstruction. It appears in more explicit form as a belief that is shared by Hugh Lehman (1977), who has written an interesting philosophical essay on the meaning of understanding in mathematics. As a clue to what may be problematic about Poincaré's remarks, consider the following comment by Lehman:

> In my mind, the educational aim of understanding mathematics does not include the aim of being a creative mathematician. . . . A creative mathematician can formulate mathematically new interesting or important theorems, or can prove theorems not yet proven, or can devise more adequate kinds of proofs. . . . Understanding mathematics . . . may be a necessary condition for being a creative mathematician. But it is possible to understand mathematics . . . and not be a creative mathematician. (p. 111)

What do Lehman and Poincaré have in common? One salient feature for our purposes is that they make a distinction between "merely understanding" and being able to operate creatively.

While one might readily admit that people who are neophytes may not be in a position to contribute to the forefront of knowledge, it does not follow that they thereby do not operate creatively. In other words, if we take seriously the points of view expressed in the philosophical arguments of the previous subsection, then something much more is required in order to gain knowledge or to acquire an understanding of the world than merely to "follow an argument." As both Dewey and Popper come closest to pointing out in the previous section, it is in the act of trying to *negate* the bits and pieces of an argument or a demonstration that one comes to acquire an understanding of what is under investigation.[8]

In fact, what is missing in both Lehman and Poincaré's arguments is an awareness that the act of "what-if-notting" a proposition or an argument is itself creative in that it requires "tweaking" something that was not salient or that was presented in a nonproblematic fashion.

Although it may be true that the sort of "tweaking" necessary in order to make a scholarly contribution to a field requires an enormous overview of the field, something the neophyte lacks, it does not follow that anything short of a significant contribution to a field is an uncreative act. We turn to a rather simple example of the relationship between "tweaking" and understanding in order to make the point.

On Understanding: An Illustration

Several years ago, while teaching a course to prospective teachers, I was introducing the concept of prime number in systems other than the standard system of natural numbers, $N = \{1, 2, 3, 4, 5, \ldots\}$. I recalled for the students that within the standard system, we usually define a *prime number* as a number with exactly two different divisors. I reminded them that we consider 5 to be a prime because 1 and 5 are its only divisors. 4 is not prime because it has three divisors: 1, 2, and 4.

Believing that they already understood the concept of prime number in N, and hoping to extend that concept for the purpose of examining properties of an entirely different system, I suggested that they explore E = {1, 2, 4, 6, 8, 10, 12, . . .} . In order to get started, I asked them to use the same definition of prime as in N and to list the primes in E.

They all agreed that 2 was a prime since it had exactly two divisors (1 and 2); that 4 was not prime since it had more than two divisors (1, 2, and 4). When exploring the status of 6, most of the students agreed that 6 was also not prime since it had four divisors: 1, 2, 3, and 6. This reaction raised an intriguing question for me. What was it that they were holding on to as the concept of divisor in the original set of natural numbers? I therefore asked them to reexamine whether or not 5 was a prime in N. They all agreed that it was since it had only *two* divisors: 1 and 5.

When I goaded them to reconsider the issue of primeness by claiming that 2 also divided 5, most of them argued that I was wrong. Why? They said that 2 did not divide 5 "evenly." I asked them what that meant and they said that there was a "remainder" when 2 divided 5. I asked them what was wrong with "remainders" and many of the students claimed that there could not be a remainder. "What was wrong with remainders," I asked them again. They just could not be permitted as possible quotients, they told me. Why not? Though several students began to see the problem of remainders in a different light, a number of them persisted in excluding them for no reason other than "they weren't allowed."

Knowing what I know now about the pedagogical power of errors and of allowing them room to persist, and similarly about the value of allowing history to unravel at its own pace, I might have permitted this sort of conversation to persist over a long period of time. After all, some students who focused on the question of whether or not 6 was prime in E argued that it was indeed prime since when you tried to divide it by other numbers (like 2), the problem was not that the "answer" was "a fraction" or a remainder, but rather that the other divisor (3, in this case) did not itself belong to the set E.

Some students came to accept that in order for a number to be eligible as a divisor, the "quotient" also has to belong to the set. They then argued that 10, 14, 18, . . . were all prime in E. The group then began to explore properties about prime numbers in E that electrified the atmosphere. They were coming up with analogies and breakdowns between the systems N and E.

Others who rejected the concept of divides as merely excluding remainders were reluctant to engage in that exploration, although several of them joined the other group once they saw some fascinating anomalies, e.g., a breakdown of the fundamental theorem of arithmetic—that every number can be expressed as product of primes uniquely, except for order of factors.

What is there to observe about "understanding" in this episode? I suppose good arguments can be made for the point of view expressed by those who were recalcitrant as well as those who were adventurous. In fact, in retrospect, it would have been interesting to interrupt the unit and to return to similar issues regarding the actual history of negative numbers (called "numeri ficti" or "fictitious numbers"). We could have wondered why there was so much resistance to the creation of these new creatures.

Leaving such pedagogical ploys aside, however, it is worth imagining why so many of them were unwilling to explore the concept of divides in a more general light than that of forbidden remainders. What sort of understanding was missing? It is not only that many of them were reluctant to explore the concept of "primeness" in this new system, but rather that they seemed to have had a fragile notion of what it meant to be prime in the system that was supposedly quite comfortable for them. They could do all sorts of things with prime numbers and in fact were aware of some interesting properties in this system. In some important sense, however, they did not understand what a prime number is even in the set of natural numbers, by virtue of the fact that they did not see the idea as something to be captured in "flight" rather than in its "essence."

What kind of "flight" could have caught the fancy of those who were not capable of seeing "prime" in another domain as anything but literally the same as prime in N? We will explore that question in the remainder of this section, and will reexamine the issue in the following one at which point we will discuss some educational problems with the concept and use of definition.

Recall the definition of prime as it is usually given: *A number is prime if it has exactly two different divisors.* If we perform a What-If-Not on the concept of prime number, what shall we select to vary? Though we each do our own learning, and what we select for the What-If-Not microscope is in a sense personal, I present the following as an array of possible starting points. We might, for example, notice that we have focused on the number *two*. We have said that a number to be prime must have exactly *two* different divisors. What could the alternatives be?

Looking at Table 2.1, a listing of natural numbers and their divisors, we notice that some numbers are in fact primes; they have exactly two different divisors. Others however, have one divisor; some three; some four. Notice for example that the numbers 4, 9, 25 have exactly three divisors. Is there anything special about those numbers? We might observe that they are all perfect squares. What conjectures come to mind with these observations? So far, the number 24 is a prize winner in that it has the greatest number of divisors—eight. What kind of question might you ask once you have made that observation?

We could go on making observations, coming up with conjectures, and asking questions based upon our effort to seek an alternative to the focus of the definition upon *two* divisors. However, what else might we vary? Notice that the definition focuses upon divisors of a number. How might we vary that operation?

Suppose we focus upon the addends of a number. We might observe that we can find only *one* number that has exactly one addend: i.e., the number 2 has only the number 1 as an addend. The number 3 has *two* addends (1 and 2) since 3 = 1 + 1 + 1 but also 1 + 2. For the number 4,

Table 2.1 The divisors for each natural number between 1 and 25.

Number	Divisors
1	{1}
2	{1, 2}
3	{1, 3}
4	{1, 2, 4}
5	{1, 5}
6	{1, 2, 3, 6}
7	{1, 7}
8	{1, 2, 4, 8}
9	{1, 3, 9}
10	{1, 2, 5, 10}
11	{1, 11}
12	{1, 2, 3, 4, 6, 12}
13	{1, 13}
14	{1, 2, 7, 14}
23	{1, 23}
24	{1, 2, 3, 4, 6, 8, 12, 24}
25	{1, 5, 25}

there are *three* addends (1, 2, 3) that can contribute to the sum. So 1 + 1 + 1 + 1; 1 + 1 + 2; 1 + 3 all work. We could go on creating a table for addends of numbers as we did in the case of divisors. In doing so, we could then make all sorts of observations, ask questions, and make conjectures as we did before.

Somewhere along the way, we might make the observation that to the extent that the concept of "prime" number focuses on division, we are not giving an obvious feature its due. That is, when we look at the divisors of a number, we think of them not only as coming in pairs, but each element of the pair belongs to the system itself. Thus, to claim that 5 is a divisor of the number 15 is to observe that there is some other number, 3 that has the property that their product is 15. Furthermore, both 3 and 5 are natural numbers.

It is by "noticing" that we are dealing in N that we can "tweak" the system and create something like E. Exploration of the sort described above in N makes it easier to imagine what is wrong with entertaining the possibility that 2 divides 6 in E. It is not that there is some remainder or fraction as a result of the division, but that the other element of the pair (3) does not belong to E. In fact then, 6 has no divisors in E other than 1 and 6 and it is therefore prime under the definition we adopted in N.

We might make a table of divisors for the set E, as depicted in Table 2.2.

Now we could engage in the same kind of inquiry we engaged in earlier when we constructed a table of divisors in N. Some numbers have 1, some 2, some 3, some 4 divisors. What do you observe based upon this new modification? What questions, conjectures, observations would you make?

Notice that in addition to questions that might be explored independently in each of the tables, we now have access to "crosscultural" questions. Specifically, we might ask comparative questions. We noticed, for

Table 2.2 The divisors for each number of E between 1 and 10.

Number in E	Divisors in E
1	{1}
2	{1, 2}
4	{1, 2, 4}
6	{1, 6}
8	{1, 2, 4, 8}
10	{1, 10}

example, that some squares in the set of natural numbers had three divisors. Armed with that observation, what would you be inclined to ask now?

On Understanding: The Imposition of Definitions

To stipulate that we will use some briefer expression (like *prime number*) for an otherwise long-winded one is one sort of definition used in mathematical thought. It is not the only way in which definition is used, though mathematicians and educators frequently take that to be the case.[9] For example, if I ask you to define a circle, I am usually not asking for some arbitrary mathematical formulation. Rather, I assume that you have some intuitive knowledge of what a circle is and that what you come up with will be some formulation (technical or not) that accords with that definition. Here we are asking for something that describes what it is we already believe to be the case, much as we might expect in the case of a philosophical analysis of the concept of truth or justice.

The definition *prime number* tells us that "something which has exactly two different divisors" is a bit of a mouthful and if we will be using the expression many times in some context, it makes sense to select a briefer name. It is like using a person's proper name to describe her rather than saying "the person who is five feet tall and lives on the corner of Main and Elm and is in Ms. Cantor's class in elementary school."

If a student is "given" the briefer definition, there is some presumption that the object being singled out by the definition is worth some attention (or at least was worth some attention by someone at some point).

Just as in the case of using a person's proper name, there is no guarantee that we will want to make use of that name many times and if we do not wish to do so, we might eventually discard it as something to be defined and remembered. Usually objects are defined for us because someone (a teacher, a textbook writer, a parent) has found them interesting enough to warrant referring to them many times in the future—and, by implication, because they want to distinguish those objects from others that might be like them in some important ways.

We do not just name objects for the sake of naming them. Interestingly enough, this point may not be obvious, especially for those who have not yet appreciated the fact that there is a connection between naming objects and singling them out for some purpose. I was made aware of this point a number of years ago when my three-year-old son was eating watermelon and noticed that it had pits. He asked me what they are called. I commented that they are called "pits." He became quite frustrated and pointing to a particular pit in the collection on his plate commented, "No, no I mean what's that one called"?

It took a while before I realized that he was drawing an analogy between naming of his friends and naming of watermelon pits. He believed that just as his friends had different names, there must be a reason for watermelon pits to have their own names. In fact, we do not offer students an opportunity to name and explore concepts of their own that they might eventually discard for a variety of reasons.

It is actually an interesting insight for a student to come to realize that there is nothing intrinsic about selecting some objects to be named and allowing others to remain as nameless components of a larger class. For most of us, snow is snow in the continental United States, and though we have a number of distinguishing subsets of horses, we do not make the fine grained distinction among those animals that middle eastern countries create for their camels (for example). It is not only that camels are more of an object of inquiry in one culture than another, but also that the differences are important in terms of navigation and trade. Just as camels are camels for some cultures, automobiles are (or at least were until recently) probably nothing more than automobiles for societies in which horse back is the major means of transportation. Designations such as compact, sub-compact, mid-size, four-wheel drive, convertible, and the like may be irrelevant distinctions.

It is an important first step to provide some purpose and some overall vision of importance when someone who is knowledgeable stipulates a definition for a neophyte. When no purpose is offered, as is the case in much of mathematical exposition, then we put students in the position of receiving the definition as if the concept has been arbitrarily selected. It is difficult to imagine that arbitrary concepts are worth naming, no less exploring.

The matter is somewhat tricky, however, because the *purpose(s)* of singling out a concept—like that of *prime number*—may be essentially impossible to appreciate before a student has seen implications and uses of the concept. In a sense, it may not be possible to understand the purpose(s) of singling out a concept until at least some of its long range implications are explored. If expressed in relatively nontechnical jargon, however, establishing *purposes* may provide a rationale for exploration.

The point with regard to *understanding* is that as motivating as knowing purpose may be, it is frequently impossible to see purpose in all its detail before one has actually seen the complete picture. One may not be fully able to understand *why* a concept has been singled out at very early stages of inquiry, though striving to express purpose in relatively nontechnical ways, may enable the student to see that a definition is not arbitrary.[10]

Another point of entrée, however, bears on the concept of understanding, and it is one that we exhibited in the previous section with regard to prime numbers. Although understanding of the definition in the context of the larger picture may be difficult or impossible to convey, it is possible to push the bounds of the definition locally. By engaging in What-If-Not activities of the sort we described, at the very least the student can come to understand that the concept seems to be quite special even if it is not clear where it is headed. Thus with regard to prime numbers, though students may not appreciate their role in establishing the fundamental theorem of arithmetic (that there is essentially one way of expressing a natural number as a product of primes), they can come to understand that prime numbers are significantly different from others and that some numbers come close to being prime while others are off the beaten path. (Martin, 1970, discusses this point from the perspective of internal vs. external understanding).

Without that sort of exploration, definitions may be taken as "frozen" so that they are not easily seen as capable of revision and extension. It is most likely the lack of appreciation for such local What-If-Notting that accounted for the inability of my students to extend the concept of prime in a new domain.

What-If-Not: A Strategy

While it is possible to view What-If-Not as a mind set—a way of viewing the world that is open to seeing ideas and propositions about it as "otherwise"— it is also possible to see it as a scheme of sorts with a number of different phases or stages.

Stages
A return to the definition of prime numbers will serve to indicate how it is that the scheme might be seen in its splendor. It is a scheme that has evolved over time both in terms of its components and the subtlety with which I have come to view it. For an unveiling of some of the progression, refer to the following by my colleague Marion Walter and myself: Brown (1981b), Brown (1981c); Brown (1985); Brown (1996); Brown & Walter (1970); Brown & Walter (1977); Brown & Walter (1988); Brown & Walter (1990); Brown & Walter (1993); Walter & Brown (1969). It might be enlightening to read these pieces in sequence from 1969 to 1993.

First Stage: List Some Attributes of the "Object"
Here is an example of some of the attributes (if one is "given" the definition of prime number, for example).
1. The statement deals with numbers.
2. The statement deals with a particular set of numbers.
3. The statement deals with divisors.
4. The statement is expressible in the form "if P then Q."
5. The statement is usually intended to be a definition.
6. The statement uses the word "has" (implying ownership).
7. The statement implies knowledge of the meaning of *divisors*.
8. The statement focuses on a certain number of divisors.
9. That certain number is 2.
10. The implication is that the two are distinct and cannot be the same.

Second Stage: Ask What-If-Not? About the Attributes
Here is an example of the What-If-Not? exploration:
1. What if the statement did not deal with numbers?
2. What if the statement did not deal with that particular set of numbers?
3. What if the statement did not deal with divisors?
4. The statement is expressible in the form "if P then Q."
5. What if the statement is not intended to be a definition?

Third Stage: Imagine Alternatives
Here are some examples of imagining alternatives:
1. The statement might deal with geometric objects (like Cuisenaire Rods).
2. The statement might deal with a system of "almost even" numbers like $E = \{1, 2, 4, 6, 8, 10, 12, \ldots\}$.
3. The statement could deal with addends.
4. The statement might be expressed in the form of a question or of a command.
5. The statement might be a theorem in the system.

Fourth Stage: Ask a Question
1. What would a prime number look like if we used Cuisenaire Rods to depict it?
2. What would be a prime number in E?
3. What would a table of addends look like for any natural number?

Fifth Stage: Analyze the Question
There is much we might do here in trying to analyze the questions asked above, as we depicted in the introduction to prime numbers in the previous subsection.

Reflection on Stages
Though it is conceptually convenient to signify separate stages, it is also important to appreciate that, in practice, the stages are very much inter-

twined. For example, consider "List the attributes" in the first stage. We are listing these attributes with the intention eventually of modifying them by asking "what-if-not?" However, as we have pointed out earlier in the discussion of the parallel postulate and also in reference to cultural differences among societies, we frequently do not even *see* an attribute since its presence is taken so much for granted. That the meaning of divisor might be something to question, and perhaps to modify, was not apparent until after the issue became problematic in seeking the prime numbers in E.

The second stage—asking "what-if-not?"—is closely connected with the first and the third. It is conceivable that we might see an attribute and not be able to negate it as an essential ingredient of the concept. We could conceivably see, for example, that a statement is a definition but not be able to imagine that it could be otherwise. Such difficulty in imagining alternatives would most likely disincline us from engaging in the second stage at all in this case, if it were not made an explicit stage of its own. It is perhaps one of the most difficult, yet educationally valuable experiences to force ourselves to ask "what-if-not?" even when we cannot see very clearly alternative possibilities.

About the third stage—varying an attribute—much has to do not only with what we *notice*, but what our (even modest) intentions might be. We might, for example vary the attribute that the number of divisors is two—looking for example for numbers that have three or four divisors—in order to acquire a clearer idea about the concept of prime number. Thus, we might try to focus on the number of divisors being three for the purpose of comparing properties of such numbers with properties (perhaps not yet explored) of prime numbers. Then we might be able to find out what is special about prime numbers.

Asking a question—the fourth stage—is among the most interesting. The questions we might ask are wide open, though here is the stage at which we frequently operate as if the modification of an attribute automatically implies a question or a sort of question. If, for example, we modify the domain from N, the set of natural numbers, to that of E, the "almost" even numbers, we might be tempted to ask what the primes are in E. There is of course much that we could ask that would engage us in what we have called "cross-cultural" study. So, if we know many properties about the set of natural numbers and the primes therein, we might be inclined to explore a similar set of questions in E. We might, for example ask if there is an infinite number of primes in E.

Even so, nothing forces us to ask questions that interconnect with prior knowledge. Ironically enough, asking a question that is totally disconnected from the original system may have the effect of generating a whole new set of issues about the original system that we had not observed before doing a What-If-Not on it. For example, if we notice that there is a simple and predictable pattern of prime numbers in E, we might ask: "Is there a formula that would generate all the primes in E?" A small amount of reflection on the question reveals that we are seeking a formula that will generate all the primes in sequence. Fortified by that realization, we can return to a similar question in N and notice, perhaps for the first time, that the request to seek a generating formula in N may have been more gluttonous than we realized. More precisely, what we found difficult in N was not only seeking a formula that would yield all the primes in sequence but one that would have yielded an infinite number of primes (and only primes), even if many were excluded along the way. We will return to this issue in chapter 5 in the section "Math *Is* Funny."

The analysis of a question, stage five, frequently discloses that there is a mismatch between the level of difficulty in the original system and the derived one. It may be the case that the tools of analysis required in the derived system are so different from those in the original that we would come to wonder about what is special about one of the systems that might account for the difference. If for example, we try to figure out if there is a formula for generating all the prime numbers in the set of natural numbers, we will most likely end up in a state of mild frustration. A similar state of frustration may very well be abated in the set E. Now of course we are in a position to wonder about why that is the case.

We now take a detour in chapter 3 in order to explore the educational travesty of collapsing problem to problem solving. Having done so, we will return in chapter 4 to expand upon a number of issues regarding problem posing and problem solving, some of which we have begun to raise in this chapter.

Chapter 3

The Concept of Problem and Its Educational Fallacy

> Probably she *was* the brightest student in Nevada City High. Somewhere, sometime, somebody taught her to question everything—though it might have been a good thing if he'd also taught her to question the act of questioning.
> —Wallace Stegner, *Angle of Repose*

We return in this chapter to the perplexing issue we unearthed in our discussion in chapter 1—revealed in our use of the modified game of jeopardy. Recall that there was a strongly held belief, among participants in my philosophy colloquium, in the intertwining of problem and problem solving. Furthermore, there was a clear directionality implied in that belief. When asked why problems are valuable educationally, the answer was that they are worthwhile for a variety of reasons that depended primarily upon their use in problem solving. The scheme in figure 3.1 is suggested.

We have come across subtle variations of this paradigm. Much of the rationale for problem solving in *The Standards* and similar documents places a focus not on solutions per se but rather on *seeking* solutions and on the uses of heuristics for achieving them, as we discussed in chapter 1. Though much of our discussion of problem *posing so far* appeared to be a diversion from solving in an immediate sense, there was still the inherent implication that it is a way station which serves the same purpose. In making use of and reflecting upon heuristics and also in posing problems, there is the concept of solution that acts as a velvet hammer.

In the case of problem posing for example, we have shown how it is that making use of What-If-Not thinking is valuable in that it helps us to understand better, and sometimes prior to the actual solution, what is significant about problem solving in that domain. As we engaged in seeking variations of prime numbers for example, it became clearer why it is that prime numbers are seen as building blocks for the natural numbers.

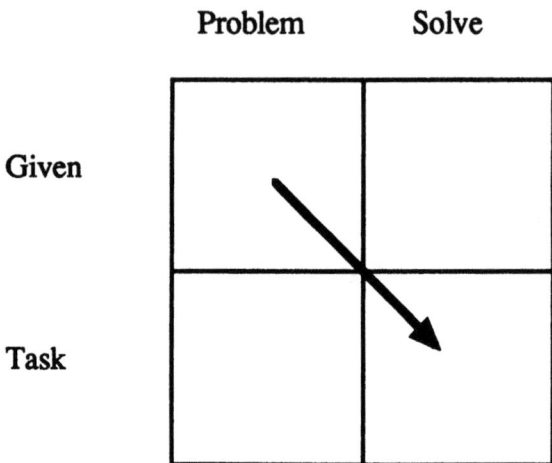

Figure 3.1 Problems traditionally seen as objects to be solved.

Further evidence of the pull of the velvet hammer is revealed in the following anecdote. I was having a conversation recently with a colleague of mine who has done some exploration in the area of problem posing. He was about to present a workshop to a group of professionals, some of whom had already heard about problem posing. He wanted my opinion about what he might do in order to disincline them from "getting hooked" prematurely (i.e., distracted by) on any of the clever problems that he had them pose during his presentation. Subsequent conversation revealed that "getting hooked" meant "trying to solve" that problem or set of problems.

What is going on here and how might we think of problems in such a way that they have the potential to connect with deeper educational and human issues, issues that stretch the bounds of what we have portrayed as humanistic in chapter 1? In preparing to do so, we shall come up with a statement of the fallacy and review the roots that seems to be responsible for the association of problem with solution in educational contexts.

The Concept of Problem: Demand and Constraints

Though problem solving is deeply entrenched not only in schooling but in all aspects of culture and though problems themselves abound in both arenas, the concept of problem itself has received relatively little analysis in educational literature. The problem of problem and its definition is well formulated by Agre (1982). He comments,

The Concept of Problem and Its Educational Fallacy   71

> Studying and solving problems has been understood widely in nineteenth-century America to be the chief means by which science, technology, philosophy, education, and democratic society progress.
>
> Learning how to solve problems has been recommended therefore as an important activity of the school. Some thinkers have even gone so far as to suggest that it is impossible to learn anything in or outside of school except as the outcome of solving problem. So what then makes any problem a problem? There exist surprisingly few sources to which one can turn for clarification about the concept of problem. (p. 121)

He points out correctly that Dewey's (1910) *How We Think* is viewed by many educators as a definitive work on the concept of problem. Agre (1982) argues, however, that "his comments which bear on the concept of problem are unsystematic and incidental to his exposition of a theory of thinking" (p. 121). In fact, much of Dewey's discussion of problem is embedded in his program to conceive of reflective thinking not as something that takes place according to some "top down" set of rules that can be accomplished in the abstract, but rather as an activity that is generated by specific events and circumstances in the real world. Furthermore, he sees thinking in even the most practical of circumstances as mirroring what goes on in scientific thought at the highest level.

For Dewey (1910), thinking begins with some difficulty, perplexity, or doubt. He sees the next step in thinking to be the formation of a plan (like the drawing up of an hypothesis in science) "or project, the entertaining of some theory which will account for the peculiarities in question, the consideration for some solution for the problem" (p. 12).

He points out that the data never supply the solution; they only suggest the plan to follow in order to reach a solution. Notice that although Dewey never actually defines what it means for something to be a problem, he assumes that the perplexity or confusion either is itself the problem or leads to the problem. More importantly for our purposes, he has established an explicit association between a problem and its possible solutions.[1]

This linkage pervades much of the philosophical literature on problems and their conceptions. In fact, though there is a more robust literature on the nature of problem than Agre acknowledges, most of that literature further entrenches the connection between problem and solution. Let us look at several other developments. Our purpose is not so much to criticize the different conceptions of problem but rather to point out the depth of the linkage and to suggest eventually that its entrenchment may be responsible for limiting our educational vision of the potential uses of problems.

Agre himself offers an interesting beginning. He sees that a problem must satisfy four conditions: (1) consciousness, (2) undesirability, (3) difficulty, and (4) solvability. Each of these conditions is elaborated on in such a way as to temper what might otherwise be a naive connection between the condition and the problem. For example, with regard to (1), it is not necessary that a person who "has a problem" necessarily be aware of it. An educator who develops curriculum in the area of problem solving may be totally unaware of the fact that she does not really understand what a problem is. For that person, even though we would not say that she is aware of the problem, we would say that she "has" a problem. We would require that there be a *someone* who is bemoaning the fact that such an educator has a problem, be conscious of the fact that it is a problem.

Similarly condition (2), *undesirability*, needs to be appropriately qualified so that the concept of problem covers not only such examples as someone having a dreaded disease, but also problems of the sort that interest us in this book: problems in mathematics or philosophy. In what sense do these problems describe something "undesirable"? It would be hard to justify an educational program that was designed to offer a regular diet of something that is undesirable.

Acknowledging that the concept of undesirability may be a bit difficult to describe, Agre points out that he merely means something that is the opposite of desirable. In that sense problems in mathematics or philosophy are undesirable in that they threaten our understanding of a situation. To have a problem in philosophy, for example, indicates that one's theoretical view is incomplete. One's world is not whole.

Condition (3), *difficulty* is an interesting one. Exactly where the difficulty lies is open. A problem may be difficult because it is difficult to formulate; it may be difficult because, though it is easy to formulate, it is hard to solve. If there is no difficulty, however, one would be inclined to claim that it really is not a problem at all.

His concept of *solvability*, condition (4), also requires some further clarification. It may be that a specific problem is not solvable. It may even be the case (and he thinks, incorrectly, that mathematics is an exception to this prospect) that one cannot tell beforehand whether or not a problem is solvable. What is needed then to call it a problem, according to Agre, is that the object being called a problem bears enough of a resemblance to other circumstances that were somewhat similar to it that in fact did yield solutions.

Just as problem and solution are linked in some way(s), so the analogous pair "question and answer" are linked. There was a wonderful interdisciplinary journal entitled *Questioning Exchange* that was published in the late 1980s for only a couple of years. Edited by James Dillon, it explored some of the logical links between problem and question and also investigated some interesting psychological and educational matters about questions themselves. A number of essays by Dillon and others published about that time both in that journal and elsewhere reveal unexpected depth in matters that may appear at first blush to be humdrum.

For example, Dillon (1986) wonders why it is that the asking of questions in an open forum before one's colleagues is so threatening to the questioner. Most people would conclude that the issue is not at all opaque. They would say that one is admitting ignorance before a group of peers. What, however, should be so threatening about acknowledging one's ignorance? He correctly points out that the asking of a question not only is an admission that one may lack information or knowledge, but just as importantly, it acknowledges what it is that one already supposedly accepts. Every question has presumptions that are sometimes implicit but always there, and it is just as much what we disclose about our beliefs as it is the ignorance that we convey that creates something of a threat. Recall, for example, the huge presumption displayed by the small word "how" that we discussed in chapter 2 with regard to attempted proofs of the parallel postulate.

The logic of this psychological dimension is well captured in one of the earliest articles on the nature of question by Felix Cohen (1929). For Cohen, a question is essentially a propositional form or function. Choosing mathematics and logic, he points out that "What is the sum of 3 and 5?" is a question which for all intents and purposes is like the sentence:

$$x = 3 + 5$$

What is obvious is that the above sentence is neither true nor false as it stands. It is, however, associated with propositions that do have truth value. Such propositions are answers to questions. Thus: $8 = 3 + 5$ is an answer to a question of the sort we discussed above. It is not the only answer. Another answer might be expressed in existential form: Thus, "There exists a number x so that $x = 3 + 5$" is an answer to the question as well. He allows for the possibility that "There does not exist an x such that $x = 3 + 5$" is also an answer to the question. It happens to be an answer that is false. For a question to be *significant*, Cohen requires that it be presented in a propositional form that has exactly one true proposition associated with it. Consider, for example, the following question:

Who discovered America in 1491 (sic, regarding political correctness)?

Questions of that sort lack *significance* because they are *invalid* in the same sense that our earlier question "How can we appropriately punish our teen-age children so that they will grow up to be good citizens?" might be an invalid question. That is, there is *not* at least one true proposition associated with it.

Similarly questions like: "Who did what to whom and when?" is a question that he would deem *insignificant* (in terms of what it signifies, not its importance) on the grounds not that it is *invalid*, but rather that it is *indeterminate* since there are so many truthful propositions associated with it that we essentially do not know what is being asked, despite the fact that it has a surface appearance of a question.

We might want to encourage students to ask questions that lack *significance* in the sense described above, and we might also wish to be softer in our requirement to focus on truth and falsity of associated propositions as the mainstay of legitimate questions. Nevertheless and despite the fact that it is dated in some important ways, such an analysis is helpful. Not only does Cohen's analysis lead us to seek analogies between question/answer and problem/solution, but it serves to highlight once more that questions contain not only an admission of ignorance, but, by virtue of the fact that they have associated propositions, they *assume knowledge* as well.

The acknowledgment of critical information as an important ingredient of problem in relation to solution is further explored by Nickles (1981). He lays down a number of requirements that anything called a problem (with a focus on scientific problems, though his analysis would not eliminate much of what we would call nonscientific problems) need satisfy in coming up with his final conception of problem.

Among them are (1) logical and conceptual ones (such as "problems are sometimes solved, i.e., their solutions are discovered, thus making inquiry possible"—an innocent sounding concept if not for questions raised by the Meno Dilemma.); (2) evidence that problems have conceptual depth (such as "The discovery process typically is structured in time rather than being a momentary psychological experience of the solution popping into someone's head"); (3) evidence that conceptual constraints belong to the problem and cannot be removed to the background (as in the realization that "some problems are deeper than others").

Based on this elaborate scheme, he criticizes a number of competing notions of problem, especially those from the school of logical positivism, and comes up with a rather simply stated conception of problem. He states, "What, then, are problems? My short answer is that a problem consists of *all* the conditions or constraints on the solution plus the demand that the solution (an object satisfying the constraints) be found. . . . Specific types of problems will, of course, possess special features. But what else could a problem in general include than the constraints plus the demand?" (p. 109).

Notice that this definition of problem, "all the conditions or constraints on the solution plus the demand that the solution be found," both softens some of Agre's categories and in addition sharpens an aspect of Agre's analysis that was stated in weaker terms. For example, it is not essential that something be "difficult" to be a problem. As with Cohen's distinction between significant and insignificant problems, we can have problems that are difficult and problems that are easy.

What Nickles does make clear, that Agre mentions as only one option among many with regard to the concept of solution, is that in order for something to be a problem, there must be some demand for a solution built in. That demand may be subtle, but if there is no request for a solution, then a problem may be like a painting or a daydream. It does not require any action by anyone. This demand for a solution is critical from our point of view since, though it is rarely expressed in any explicit sense by advocates of a problem solving curriculum, it is the implicit acceptance of that point of view that accounts for the fallacy we will explore shortly.

Before doing so, however, it is worth acknowledging that this brief definition has a lot more that is implicit than appears on the surface. For example, there is a fascinating debate in the philosophy of science about the role of the history of a problem in relation to the problem. Some would claim that the history of a problem, regardless of how interesting it may be, is not necessarily part of the problem itself. Others would claim that the history of the problem is part of the definition of the problem in that discounting the history of the problem, when it came about, what prior efforts were made in solving it, when it may have been abandoned as a problem, would violate its status as a problem (Hattiangadi, 1978; Sintonen, 1985).

Another aspect of the debate centers on the question of whether or not there are in fact unsolvable problems. While the question is interesting, what is more intriguing is the set of issues that surface in an effort to have the question make sense. Moritz Schlick (1935), clarifying the difference between questions (which are cousins of problems) that are answerable in practice or in principle, concludes, "There are many questions which are empirically impossible to answer, but not a single real question for which it would be logically impossible to find a solution" (p. 25). Without reviewing the details of his argument, it is built into his concept of question that all questions are answerable in principle.[2]

Rapaport (1982), looking at some of the same issues that Schlick does, elaborates upon the concept of problem by making it clear that all problems are theory laden. At first one would think that since theories are frequently in conflict with each other, this point of view would argue for most problems being fundamentally unsolvable. Any problem depends upon the assumptions that are built in, and the assumptions differ depending upon one's theoretical orientation. This is a position that is similar to the *constraints* that Nickels spoke of in his definition of problem. Yet, he moves beyond a context-bound position ("it *depends*") and comes to the conclusion that there can be solutions without waffling over which is "the correct theory" for—borrowing from a scheme developed by Perry (1970)—part of being human is *committing* oneself to a point of view. He uses this insight to shift the original question of whether or not there are unsolvable problems and uses it to propose a positive answer to the question of what constitutes progress in fields such as philosophy.

The Fallacy

We have meandered a bit in reviewing various conceptions of problem. These conceptions have come mostly from philosophy and in particular from the philosophy of science. Despite the interesting side paths that are embedded in these debates, there does emerge a rather strong conclusion: Problems are linked to solutions. More precisely, the *concept* of problem is linked *to the concept* of solution. Deliberation over whether or not every problem is solvable, while it may appear to provide a voice of skepticism about the connection, in fact further entrenches an inclination to seek the sense(s) in which there is an affirmative answer to the question of linkage.

The nature of the fallacy is primarily a logical one, and one we shall sharpen in the following section. It derives from an ancient view of the potential of objects in the world, a view that we shall explore in the section, "Roots of the Fallacy." This fallacy accounts for our myopia regarding the educational potential of problems. To realize that the fallacy exists, however, does not automatically suggest methods for its rectification. For that we need both vision and an alternative perspective. Part of it has

The Concept of Problem and Its Educational Fallacy 77

been provided in chapter 2, prior to our analysis. Additional perspective is provided in chapter 4.

Consider the question: How can a problem(s) be used for educational purposes? Where do we look for an answer to that question? It is in trying to come to terms with these questions that the fallacy emerges and eventually cascades. It is the linking of "what a problem is" with "how it can be used" that prematurely draws us to view "solution" as a critical component of problem for educational purposes.

As we have said earlier, the ways in which the connection is made have taken on a new complexion in recent years. *The Standards* and related documents have softened the actual solution as the prize and have focused more heavily upon the process. Hence, we hear a lot about *heuristics* for problem solving. There is much talk of journal writing in mathematics classes and a great deal of rhetoric that encourages students to record the difficulties they have and the triumphs they achieve in an effort to solve problems. Such procedures, together with many of the activities associated with problem posing, are done, however in an effort to nail down tasks associated with problem solving, though frequently done with subtlety.

We have encountered the fallacy in a number of compelling but imprecise ways throughout the text. The analysis of the concept of problem in the above section enables us to offer a more careful rendition of it.

Sharper Statement of the Fallacy

Though there are a number of alternative conceptions of *problem* and though there are many different conditions that seem to be part of its definition, the driving logical/philosophical force is captured most sharply by Nickles. Recall that according to him, "a problem consists of all the conditions or constraints on the solution plus the demand that the solution be found." It is Nickles' clear statement of the *demand* condition linked with the desire for a solution, that contributes to the educational fallacy. In other words, it would be possible to view a problem and to be aware of the fact that as part of its condition it has a built-in demand for solution without feeling compelled to accede to that demand. In a nutshell, it would be possible to view a problem as an interesting situation that can be used for a variety of purposes without responding to the (sometimes implicit) statement of demand.

The educational fallacy that flows from the pair, demand/solution, comprises what most people intuitively see as necessary ingredients of the *concept* of problem. Even if they do not so articulate them explicitly, there is the inclination to adopt the point of view we depicted schematically in Figure 3.1.

There are in essence two fallacies rather than one. One is a kind of metafallacy, something of a logical sort; the other is an educational fallacy that flows from the metafallacy.

We have been somewhat ambiguous in distinguishing between the two of them. Part of the reason is that they are intertwined, and it is difficult in practice to separate them. We shall continue to allow for a bit of ambiguity, but for conceptual clarity, we make the distinction.

> *Meta Fallacy*: Once we are clear about what we mean by problem, and how it relates to constraints, demand and solution, then an educational agenda flows from that conception.
>
> *Educational Fallacy*: The educational agenda is roughly depicted in the sketch of Figure 3.1.

It is necessary to take "roughly" in the above remark seriously, since as we have pointed out, activities such as problem posing are acknowledged to be educationally worthwhile. Such activity, however, is instrumental frequently in furthering an ability to solve specific problems or to develop problem solving skills more generally. A clearer statement of what is involved in Figure 3.1 is:

> *Given any problem, in order for it to be used for educational purposes, it is necessary that the goal ultimately be directed towards a possibly wide variety of efforts of students to solve that problem.*

Roots of the Fallacy

The fallacy is rooted in an Aristotelian view of objects having intrinsic properties. There is nothing inherent in the concept of problem that requires that we select its *definition* as its educational purpose. Scheffler (1985) clearly spells out the shortcoming of this point of view with regard to the concept of an acorn. Paraphrasing Aristotle's position he says, "The acorn, potentially an oak, is moved by its potential oakhood toward realizing the final state comprising its natural end, the features of which define the essence of its species, and explain the growth stages of its members" (p. 42). He then points out the difficulties with that scheme:

> The difficulties of this scheme, from a contemporary point of view, are notorious. The whole notion of species as fixed has been supplanted, through the rise of evolutionary theory, by the notion of species as alterable; the idea of a perma-

nent nature with an enduring essence has simply been surrendered. Independently of this shift in biological thinking, there are difficulties in understanding and applying the basic concepts. What is an essence anyway? An acorn is not only the seed of an oak but a morsel prized by squirrels. Why is the one property singled out as natural and explanatory, in preference to the other? Is the first more predictive than the second? But more acorns are eaten by squirrels than grow into oaks. True, if an acorn is properly nurtured, and protected from squirrels, etc., it will grow into an oak but, equally, if it is properly placed in the vicinity of hungry squirrels, it will, on the contrary, be eaten by a squirrel. Reference to *essential oakhood* will in any case, not predict the outcome, and is thus of no use in explaining the acorn's development. (pp. 42–43, emphasis in original)

How we should make use of problems for educational purposes is not something we can determine by adopting anyone's conception of problem alone. Not only does it require an inquiry into why we educate, but it is something that cannot necessarily be read by deeply analyzing what must have been in the minds of "experts" who claim to know what a problem "really is" and, by implication, how it might be used for educational purposes.

Dennett (1990), whose major intellectual focus is on viewing *intentionality* in an holistic way, offers some fascinating arguments for why it is that we need to be extremely cautious in asking of artifacts what purpose they serve or of authors of texts what "they really mean" by the stories they write or in asking of a field such as biology why an organism has been selected to survive as it did. He speaks of such undeserved confidence as "the intentional fallacy," a phrase created originally in the field of literary criticism, and comments,

Occasionally, an artifact loses its original function and takes on a new one. People buy old-fashioned sad-irons not to iron their clothes with, but to use as bookends or doorstops; a handsome jam pot can become a pencil holder, and lobster traps get recycled as outdoor planters. The fact is that sad-irons are much better as bookends than they are at ironing clothes—at least compared to the competition today. And a DEC 10 mainframe computer today makes a nifty heavy-duty anchor for a large boat mooting. No artifact is immune from such appropriation, and however clearly its original purpose may be read from its current form, its new purpose may be related to that original purpose by mere historic accident—the fellow who owned the obsolete mainframe needed an anchor badly, and opportunistically pressed it into service.

Inventors of artifacts are no more immune to confusion than authors of texts. It is possible that someone setting out with every intention of creating a new kind of alarm clock succeeds, in spite of himself, in creating something that can *charitably* be described as merely a new kind of paperweight. Consider how the Intentional Fallacy looks when applied to artifacts: the inventor is not the final arbiter of what an artifact is, or is for; the *users* decide that. The inventor is just another user, only circumstantially and defeasibly privileged in his knowledge of the functions and uses of his device. (p. 184, emphasis in original)

With the full realization that it is human intention that defines how problems will be used in educational contexts, and with the understanding that a humanistic orientation as we explored it in chapter 1 is both educationally defensible and unrealized in mathematics education, we turn in chapter 4 to ways of reconstruing how and what we investigate about problems.

In making that shift in orientation, we find the following conception of problem, coming from Richard Wertime (1979), an author with a phenomenological orientation, inspiring. It is the only essay concerned with problem solving that appears in a book devoted to an analysis of the educational role of cognitive science that is less concerned with technical matters and more with what it means to be a person in relation to problems and problem solving. What follows in chapter 4 may be viewed as an attempt to fully appreciate the notion of problem as intimately connected with human events. It is a theme that is picked up in chapters 5 and 6 as well, though there we are more frontally concerned with "applications" of mathematics and its connections with "the real world." Wertime (1979) comments:

> A problem is . . . a project for the future we commit ourselves to by an act of will. This means by implication that a problem entails risks, since all future - projects—to use Hannah Arendt's term—involve uncertainty. There is . . . an essentially promissory dimension to the act of facing a problem. A problem is not an entity which has existence independent of a person. . . . They involve a significant act of self-surrender which can seriously jeopardize the individual's sense of self. (pp. 192–193).

Though it lacks the precision of Nickels, Agre, Rapaport and others, it does inspire us to look in other places for how problems might be used than those that have been advocated by documents such as *The Standards*.

Chapter 4

Problem Posing/Solving in a Humanistic Light: Softening the Fallacy

> I was walking across a bridge one day, and I saw a man standing on the edge, about to jump off. So I ran over and said "Stop! don't do it!" "Why shouldn't I?" he said. I said, "Well, there's so much to live for!" He said, "Like what?" I said, "Well. . . are you religious or atheist?" He said, "Religious." I said, "Me too! Are you Christian or Buddhist?" He said, "Christian." I said, "Me too! Are you Catholic or Protestant?" He said, "Protestant." I said, "Me too! Are you Episcopalian or Baptist?" He said, "Baptist!" I said, "Wow! Me too! Are you Baptist Church of God or Baptist Church of the Lord?" He said, "Baptist Church of God!" I said , "Me too! Are you original Baptist Church of God, or are you reformed Baptist Church of God?" He said, "Reformed Baptist Church of God!" I said, "Me too! Are you Reformed Baptist Church of God, reformation of 1879, or Reformed Baptist Church of God, reformation of 1915?" He said, "Reformed Baptist Church of God, reformation of 1915!" I said, "Die, heretic scum," and pushed him off.
> —Emo Phillips

In seeking ways of viewing problems that do not focus primarily on their solution, we will entertain some ways of portraying problems that sever the bond, some ways that redefine the nature of the bond, and some that actually strengthen it. Though the latter may appear to be a bit ironic, it is worth stressing that it was in the act of exploring their independence that new ways of seeing potential connections arose.

What is wrong-headed about seeing the relationship between problem and problem solving that is depicted in chapter 3? First of all, when collections of problems are presented primarily as "demands" to seek their solution, and are isolated from context, intention, and history, the activity tends to trivialize the human drama that is associated with inquiry. As we have discussed earlier, there is nothing that inheres in objects (including problems) that defines their intentions, educational and otherwise. Intention is a human quality that motivates our inquiry, and to see intention either as irrelevant or as unidimensionally defined is to rob inquiry of its most human possessions.

Second, as we began to argue in chapter 2, when any problem is solved without engaging in the posing of other problems, there are aspects of understanding that are forfeited. Furthermore, as we hypothesized in that chapter, the disinclination to pose problems in the sense of challenging the "given" frequently results in a type of rigidity that interferes with future exploration.

Third, viewing problems exclusively as invitations for their solution provides an inaccurate picture of how the discipline itself grows. We spoke earlier of Lakatos' view of progress as connected with production of counterexamples rather than problem solving as normally conceived. The analysis of progress, however, is considerably more interesting than that. We will indicate how aspects of a field move forward not by solving new problems exclusively, but by reconstruing what the field is about. How we should construe a field is not something that "pops out" of a set of solutions to problems, but rather requires a kind of insight that connects in a more global way with problems that have or have not been solved.

Finally (well, almost finally), problems can be used for the purpose of shedding light on the idiosyncratic ways we each view the world. In addition, the fact that some issues are explicitly singled out as problems and

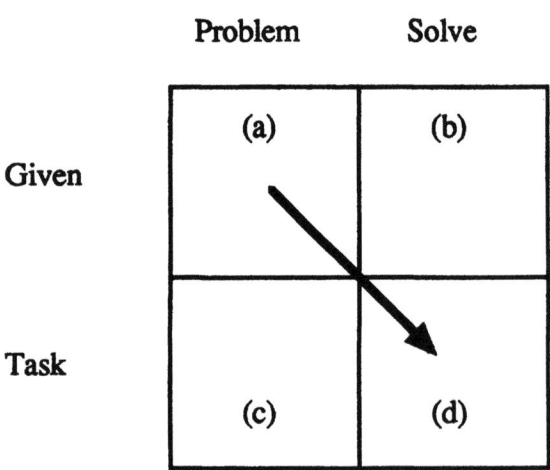

Figure 4.1 Readying the traditional view of problem solving for criticism.

others not, signals what it is that society takes seriously. Furthermore, our reaction to problems enables us to see in a new and sharper light what sort of things we fear and what we revere, what it means to have a mind and how other minds function.

As we mentioned above, some of these ways of relating to problems will in fact invoke the concept of solution. However, when we turn from the "meaning" of problem to "educational purpose," there is at the very least a flipping of "solution" from foreground to background position. The concept of solution may be present but, if so, it will transform the global focus we depicted in chapter 3. Thus, educational purposes and solutions of problems become more reverberational.

Since we will be rectifying and elaborating upon the scheme that is associated with the fallacy analyzed in chapter 3, we repeat the diagram of Figure 3.1, but this time, we designate the boxes of each quadrant in order to ready the scheme for elaboration. See Figure 4.1.

Scope and Intentionality

We began a discussion of intentionality with regard to definition in chapter 2. We elaborate upon that issue here as we combine intentionality with scope. A concern with intentionality and scope enables us to move from an isolated array of problems-to-be-solved towards a focus on problems writ large. The new focus enables us to return to the nature of the problem-to-be-solved in a new light. Once the newly conceived problem has a broader scope, the question of what can be done with that problem is once again not necessarily determined. Beginning with scope and intentionality, the rest of this chapter will investigate the sorts of humanistic pursuits that might be associated with problems as given.

Below is an anecdote that conveys a great deal about the limited scope of problem solving in much of the curriculum.

> A number of years ago my son, Jordan, wanted to talk to me about "the ambiguous case" in trigonometry—those circumstances under which a triangle is determined by two angles and a side not included between the angles. I began the conversation by asking him to recall that in his previous geometry course he had investigated those circumstances under which a triangle is determined. Jordan was quite puzzled and he told me that he had never studied such a thing. They supposedly had never investigated the determination of a triangle. Rather, he told me that they had proven things about two triangles being congruent if A.S.A. equals A.S.A. and so forth. He reminded me that they did a great deal of problem solving using such schemes and that sometimes it was difficult to locate the corresponding pieces and that it was fun to find such hidden information.[1]

What has taken place here? Jordan was well accustomed to experiencing mathematics as *problem solving*. What he had rarely been invited to see as problematic, however, is the scope within which such problem, solving takes place. Many activities had been constructed which formally and informally introduced the concept of congruency for triangles. Many of them involve the student in clever problem solving acts and in reflections upon that problem solving. What most programs overlook, however, is the need to include some sense of scope—a realization that a focus on triangles is not a first step in inquiry about congruence, but that it is essentially a last one. It is a step that had to come about historically and logically in the investigation of congruency for more complicated rectilinear figures and requires ultimately an appreciation for the fact that triangulation reduces a complex problem to a simpler, more manageable one.

What is needed in order for students to appreciate a larger purpose for focusing on triangles is an inversion of the overly worn maxims such as "in learning it makes sense to proceed from the simple to the complex," or that one ought to be taught so that ideas unravel from the concrete to the abstract.

As we discussed with regard to purpose in relation to definition and understanding, however, the issue is quite complicated since it is not possible for a number of good reasons to always involve students in seeking *purpose* and *intention*. Labor pains may need to be experienced in order to appreciate the relationship between the big and the small picture. It may not be possible, for example, for a teacher to explain why algebraic thinking is valuable at the point at which students are just beginning to acquire an understanding of its abstract nature and its relationship to more concrete ways of thinking.

Furthermore, when too much of the story is told beforehand, much of the surprise is depleted from the subsequent adventure. To convey when one is being introduced to the concept of irrational numbers, for example, that not only are there an infinite number of such creatures, but there are an uncountable number, destroys the reasonable expectation that not even one of them exists.

What is needed, however, is enough sensitivity to the concept of scope so that teachers can entertain the possibility that the big picture can sometimes be conveyed even before the details are filled in. It is worth taking into consideration that providing the most imaginative of problems to solve will not necessarily provide a picture of the scope of that problem solving activity. It takes a great deal of imagination to find a proper balance between problem solving in the small (e.g., proving congruencies for triangles) and devising a curriculum that will sensitize students to scope and intentionality through problem solving in the large, as in creating problems about congruency for complicated figures before learning anything about triangles. It will take even greater imagination to find a balance between problem solving and other modes of communicating to students what the purpose might be for including much of the curriculum.[2]

Just as important as "conveying" scope to students, however, is the value of investigating how students see both scope and intentionality in problems as they try to unravel what is "figure" and what is "ground." I see this concern as a major contribution of constructivism. One of the most enlightening experiences I had with regard to students' understanding of scope occurred several years ago when I spent a semester observing a seventh-grade accelerated algebra class. The teacher had devoted considerable time teaching the concept of "unknown." She had done so imaginatively and was interested in having the students not only manipulate expressions, but understand what a variable was all about. After several weeks had passed, I interviewed Debby, whom the teacher described as the best student she ever had. She seemed to have the distinction clear between a symbol and what it stands for, between an expression and a solution to an equation, about ways to justify movement from one expression to another, and so forth. I then asked her the question: What do you expect to learn next semester when you study more algebra?

I anticipated that she might talk about more complicated expressions and equations to solve, new systems of numbers, possible applications to other nonmathematical areas, possible ways to view what they had learned from a different point of view. Instead she answered, "I will finally learn what these x's and y's really are."

It was obvious that Debby had an agenda that was quite different from her teacher's. She was telling herself a different story. As far as I can tell, she was letting me know that she hoped that someday these "unknowns" would be become "knowable." It was almost as if she was paraphrasing from the Bible—now she sees through a glass darkly; then she will see clearly.

The concept of scope and its relation to intentionality, as portrayed in the examples of Jordan and Debby, provides us with a spotlight on an important aspect of meaning, a spotlight that is not always made brighter by having students *solve* problems per se. Meaning means many things. As we saw in the discussion of theoretical issues related to What-If-Not thinking, some philosophers associate meaning with the potential for a statement to be falsified or shown to be true. Others are concerned with some justification of a statement. To treat an arithmetical calculation as *meaningful* is to acknowledge that the concept is not arbitrary and that it has a home. More than that, it has a home that is understandable in the sense that it connects with an already existing structure.[3] To be taught subtraction exclusively as a collection of rules for "borrowing" is to invite a mechanical responses that could never have been devised by the student. While it might be the case that students could not have come up with specific algorithms on their own, when a meaningful concept of subtraction and of place value is available, such procedures have a coherence that are capable of being *understood* even if they have not been created by the student.

The concept of purpose, however, provides another dimension of what we seek when we claim that we want something to be meaningful. We frequently confuse two notions of meaning—being able to justify and seeing scope. The use of concrete materials is surely helpful for many reasons in introducing new ideas. To make something more concrete, however, does not always clarify why it is that the idea is worth investigating. Use of chips or of an abacus may be helpful for the purpose of illuminating why we use some of the algorithms we do in performing operations on natural numbers. They do not, however, in and of themselves make clear why we would ever have been drawn to create such schemes. What is needed is a set of experiences to make clear why it is that primitive ways of characterizing numbers—such as using many stones to convey how many horses there are in the village—has its shortcomings. That is, place value was created for a purpose and that purpose is not revealed in learning to handle place value problems in a "meaningful" way. We do not always appreciate that students who ask "why" something is the case, are sometimes asking not for a reason in the sense of a logical or scientific explanation, but for a purpose served by an invitation to inquire. Brown (1982) discusses the nature of this confusion in relation to a student's poignant moral dilemma. In doing so, he shows how a teacher's courage to confront purpose in ethical rather than strictly scientific terms has the power to rejuvenate student interest and commitment.

The kind of diagnosis I began to offer in the case of Jordan and Debby is not meant as a "put-down" of either of them. While it is helpful for us as teachers to see what we are doing in some broader context than the specific problem or set of problems that are part of a unit, there is nothing that is written in stone about purpose or set of purposes. I continue to marvel at how readily my students see worlds that I had never imagined as they supposedly "misperceive" what I saw as the central focus. Pursuing their supposed misperceptions has often led to the exploration of valuable territory that no one in the world may have seen before. If we teach with the expectation that errors, questions from left field, and confusions have generative power, then we can appreciate that even supposedly naive students are capable of changing the way we view any aspect of knowledge or experience.[4]

It is tempting to associate a teacher's "imposed" view of meaning, even if done with imagination, compassion, and concern for whether or not it is understood, with a *traditional* conception of education, and one which focuses on the child's emerging meaning as depicting a *progressive* mode. Dewey (1902), however, warns that this dualistic outlook masks an important commonalty. It looks as if it is subject matter that furnishes the end in one case, and the child, who is the starting point in the other. In fact, he points out that the child-centered theorists must see how it is that, "From the side of the child, it is a question of seeing how his experience already contains within itself elements—facts and truths—of just the same sort as those entering into the formulated study; and what is of more importance of how it contains within itself the attitudes, the motives and the interests which have operated in developing and organizing the subject matter to the plane which it now occupies" (p. 345)

Conversely, what must subject matter specialists acknowledge? They must see how the academic studies are themselves experiences that "embody the cumulative outcome of the efforts, the strivings, and successes of the human generation after generation" (p. 345).

In summarizing both perspectives, Dewey comments, "The facts and truths that enter into the child's present experience, and those contained in the subject-matter of studies, are the initial and final forms of one reality. To oppose one to the other is to oppose the infancy and maturity of the same growing life; it is to set the moving tendency and the final result of the same process against each other "(345).

What Dewey does here is powerful. As we suggested earlier, he does not sympathize with a romanticized progressive impulse. He does not tell us that we need to come down on the side of the child's meanings and interests. Furthermore, he is not saying that what we need to do is make some compromise in the sense of sometimes viewing teaching from the point of view of imposing a subject matter perspective and sometimes listening to the child. Rather, he is telling us that regardless of how we act as educators, we need to keep in mind what is common between these ostensibly different poles. In other words, we need to see at every turn that whatever we observe to be a subject matter perspective and whatever we observe about the meanings that a child imposes on the world, we would profit from trying to see these from an evolutionary perspective.

We would profit from asking questions of an epistemological nature like: What might this idea have looked like one hundred years ago, one

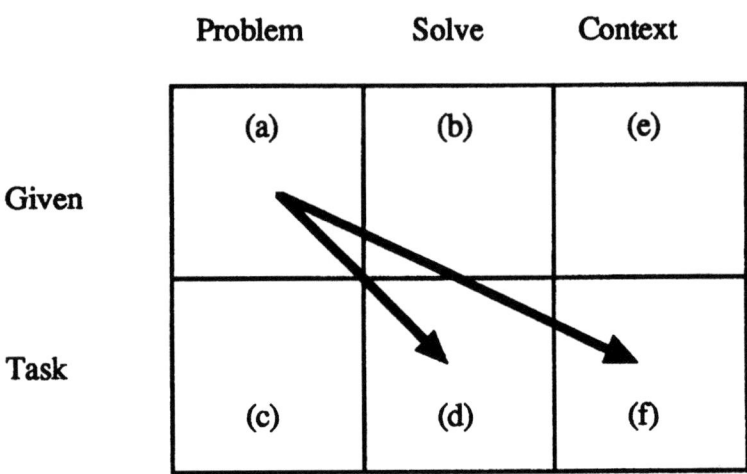

Figure 4.2 Context as a dimension of problem as given.

thousand years ago? What might have inspired us to move beyond it? What forces might have prevented us as a species from moving ahead? No matter how naive the view of the child, however, we would profit from seeing it as an early formulation of an idea that evolved into something else.

To sum up, context needs to be included and explored in a variety of ways, a dimension that has not been depicted in Figure 4.1 The context might involve scope as well as purpose of problems that are set forth for exploration. It might also involve different notions of meaning and "meaningfulness." This is just a beginning, of course, for there is a lot more that might be included in context, and other dimensions will be discussed in the following sections. Figure 4.2 depicts the ingredient of context.

The movement from box (a) to (f) may in fact be enriched by finding ways of connecting it with the movement from (a) to (d), but it is not a necessary movement. Furthermore, the kinds of problems we select for solving when box (f) is of concern, might be of a totally different kind than when it is ignored.

Problem/Situation Reverberation

What one person considers to be a problem, another may see as a situation and vice versa. Let us now look at some nonmathematical examples. Consider the following statements:

- The speed limit has been increased in many states in the United States from 55 to 65 miles per hour.

- Life expectancy has doubled in many parts of the world in the past century.
- It is possible to travel from any country in the world to almost any other country within a very short period of time (in most cases within a day).
- Women (African Americans) generally earn less salary than men (Whites) for comparable work.
- Babies are not in a position to fend for themselves.
- Half the people of the world go to bed hungry each night.

Now, what do we make of these statements? For some people, they are merely situations. For others they are problems. What is it that makes them problems? Neither the form nor the content of the statement itself provides a clue as to its status. What is needed to understand the nature of the statement is to appreciate that there is an ellipsis of sorts in that it is part of a larger context. The statements must be perceived within the context of other factual statements and also within the context of value systems. When we say that the speed limit has been increased to 65 miles per hour, we are assuming that we are referring here to vehicular travel and not pedestrians jogging. We also suspect that the change of speed limit is the result of the passage of laws that were motivated by certain interest groups and that these laws are subject to change based upon empirical evidence, e.g., percentage of accidents that result from the change, and upon what we value, e.g., the thrill of driving fast.

What we see as a problem and what we see as a situation is a critical issue, and the inability to understand that distinction frequently accounts for the difficulty of nations as well as individuals to communicate with each other.

We return to some statements/ideas that have more of a mathematical ring:

1. $x^2 + y^2 = z^2$.
2. We have an isosceles triangle.
3. Mordecai, a member of the advertising staff of the P/L Publishing Co. of NYC is enjoying his winter vacation at the Org ski resort in Orono, Maine. He is with a group of colleagues from the advertising staff of the company. His friend Pablo from the editorial staff of the same company was not able to join him when he left for Org, since the entire editorial staff was facing a deadline to edit a groundbreaking manuscript on the philosophy of mathematics education. Mordecai decides to make a long distance call to Pablo in NYC to

tell him that Org is the most magnificent ski resort he's ever attended. Pablo decides to get the boss of P/L Publishing to offer a bonus to the entire staff for completing the editing job in two days. He gets it, and they are successful in completing the task in a couple of days. He and several of his friends from the editorial staff drive to Org to join the advertising staff. Along the way to Maine, Pablo and his friends see a Mercedes ambulance that passes them along the highway coming from the other direction. They pass at a point along the road that they notice a billboard advertising free ice cream. They comment on the Mercedes ambulance because they have never seen one like it before. Though they are tempted to stop for the ice cream, they are more interested in getting to the resort quickly than in stopping.

When Pablo arrives at Org, he finds out that Mordecai and the entire advertising staff has left the ski resort because their companion Chul Hong has broken his leg and needed to see an orthopedist whose office is located in NYC next door to the P/L Publishing Co. (The staff of P/L has an HMO insurance plan that would not reimburse its members for treatment other than at the NYC office unless the medical emergency was a life and death matter—which this is not).

The next day Pablo calls Mordecai who is back at work in NYC. He asks Mordecai how they got home with Chul Hong and his broken leg. Mordecai tells him that they were able to make a special deal with an ambulance that was about to head to NYC anyway, since the driver was planning to visit his sister there.

Furthermore, Mordecai tells Pablo that he thought he passed Pablo's car along the highway heading towards them when they were in the ambulance, but it was just a passing thought. In discussing the possibility, it turns out that Mordecai noticed the free ice cream bill board sign at the same time that he recalled passing the car that might have belonged to Pablo.

When they discuss the matter further, they both recall a fact that they find astounding. That is, it turns out that each group left their respective place at precisely the same time—heading towards each other's destination.

Furthermore, since it was such a surprise to see the free ice cream sign, both groups noticed how long it took to get to their respective destination after passing it. Mordecai said it took them nine hours to get home after seeing the sign, and Pablo told him that it took four hours to reach Org after seeing the sign. Furthermore, they know that the distance between NYC and Org is 300 miles.

At what rates were each of the two vehicles traveling? (Have patience. We will see this one again).

Some may see the first example as a situation with very little to require inquiry of any sort. Some reasonable responses might be, "How nice," "I couldn't care less." However, it is possible to imagine an effort to impose something quite different on this situation. One might say, "If x, y, z, refer to integers, then I can think of some neat questions to ask?"

In the second example, a range of reactions might include something like that in (1), but in addition we might react as follows: "I also know something about the angles of the triangle,"

"Why are you telling me this?" "Do you have a specific isosceles triangle in mind?" "I wonder what would happen if I took many such equilateral triangles and tried to join them with each other in some systematic way."

The third example is particularly interesting. Despite the fact that problems (much less interesting of course) of this sort were the mainstay of problem solving in secondary school for years, they appeared to many students to be artificial or irrelevant. As importantly, they were terribly confusing. It was not only that students had difficulty figuring out what "type" (mixture, age, travel, work, and so forth) of problem they were confronted with, but more importantly, they frequently could not figure out what was "given" and how it related to what was "demanded." That is, they could not see it as a problem, at least in a mathematical sense.

To the extent that the question asked in (3) describes an insurmountable difficulty for many students, a program designed to "neutralize" problems so that they may become situations without a built-in demand might very well enable students to come to an understanding of "what the problem is." For many people, just removing the question at the end of (3) would transform a problem into a situation. For others, there may be a number of problems that remain that are implied in the description even if there is no explicit demand.

As we move from the focus of creating problems from situations, and vice versa, to that of trying to understand what motivates some people to create some situations and others to create other situations or to why some kinds of problems or situations created are more appealing to some and not to others, we begin to participate in a dialogue that enables people to reflect on what they value and how they think.

When we communicate with others about what we believe can be transformed from a problem to a situation, we tell each other what we "see" even when we may not state it explicitly. That most people do not notice that in defining prime numbers, for example, the domain of natural numbers is seen as salient (and thus capable of being transformed to some other set), tells us what they take for granted.

Such an observation is an invitation to try to figure out what it is in our own lives that we take for granted and "do not see." It becomes a significant question then to figure out what it might be that enables us to "see" better or differently. There are surely occasions upon which tragic events can become the impetus to enable us to see what was not there before. It is possible to view tragedy as a form of What-If-Not. It represents the modification of a situation in such a way that we see the situation itself differently.

In Figure 4.3, we add the dimension of situation to that of the earlier schemes. Below is a depiction of the reverberation of problem and situation. We can eliminate the demand of a problem explicitly and create something closer to a situation. Similarly, we can take a situation and impose a demand to create problems. All of this can be done in the absence of solving as a dominant force.

Of course, it is possible, as we described above, for a situation to have a number of implicit demands, even if there is nothing explicit in the statement itself. It is context and life experiences that create such an awareness. Though it is possible to move from box (g) to (h), for example, in an almost mechanical way, it is usually a movement that is undertaken with an awareness of some context. As we looked at primes in different domains, for example, we had in mind that the concept of prime itself might not be as transparent as it appeared at first glance. There was an implicit set of problems that moved us to look at the set E in relation to N. In a sense, then we were motivated to move from (g) to (h) by a desire to

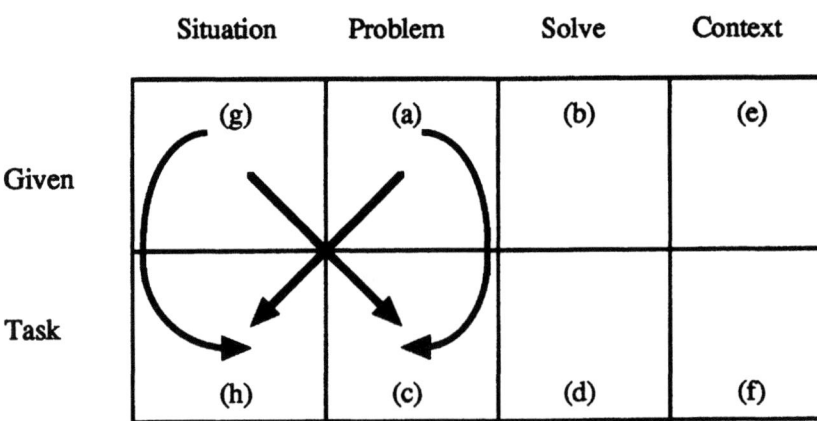

Figure 4.3 The dynamics of problem and situation.

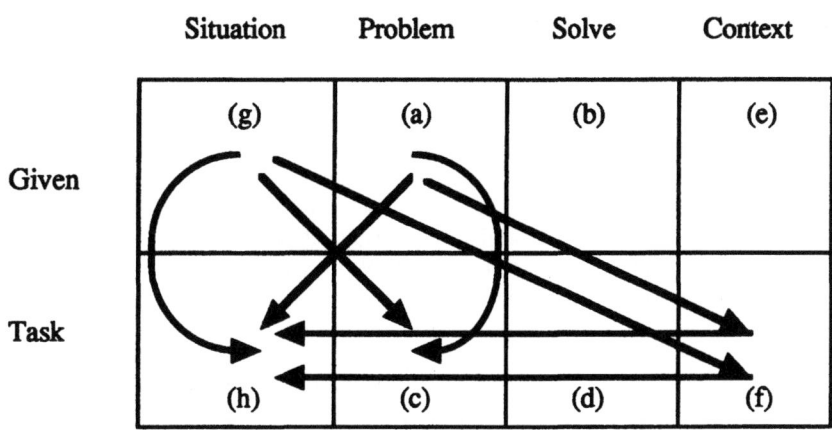

Figure 4.4 Numerous paths among situation, problem, solution, and context.

create new (g) to (h) connections. Furthermore, the admission that the concept of prime in N was not as transparent as it might have appeared at first blush, got us to incorporate context that was encapsulated in Figure 4.2. We thus have a more elaborate scheme as shown in Figure 4.4.

Thus, an awareness that the concept of prime number has some interesting contexts that we have barely appreciated is expressed in the movement from (g) to (f). This shift in turn, inspired us to move from (f) to (h)—as in creating the system we described as E. Having done so, however, we posed a new set of problems about primes, e.g., how to characterize them, essentially moving from (g) to (c).

Problem/Situation Reverberation: A Return to Progress in Mathematics

In chapter 2, we spoke of the centrality of problem solving in mathematics as a discipline. We mentioned that genius is usually defined in terms of one's ability to solve problems. Lakatos has added an interesting wrinkle to this view of progress by demonstrating the powerful role of the production of counterexamples to refute the legitimacy of alleged solutions of even relatively easily stated problems. Nevertheless, we are still left with a form of problem solving as the hallmark of progress in the field. One of the most interesting stories that supports this notion of progress can be found in the reaction of the mathematics community to the contribution of David Hilbert—whose work we discussed in chapter 1 with regard to Gödel's discoveries in the foundations of mathematics. Born in East Prussia in 1862, he was considered by many of his colleagues to be the greatest mathematician in the early decades of the twentieth century.

Hermann Weyl (1951) offers the following reason for that description. He says,

> At the International Mathematical Congress in Paris in 1900 David Hilbert, convinced that problems are the life-blood of science, formulated twenty-three unsolved problems which he expected to play an important role in the development of mathematics during the next era. How much better he predicted the future of mathematics than any politician foresaw the gifts of war and terror that the new century was about to lavish upon mankind! We mathematicians have often measured our progress by checking which of Hilbert's questions have been settled. (p. 525)

Despite the powerful tug of problem solving as the sine qua non of mathematical progress, there is a competing story to be told on the centrality of problem solving within the discipline. In chapter 2, we pointed out that it would be possible to analyze the two-thousand-year-old misperception of the status of the parallel postulate in terms of the relationship between problem and situation. It was not that a genius was needed in order to prove once and for all that Euclid's fifth postulate was provable from the others. Rather what was needed was a different sort of insight. One had to see that the problem was in some sense flawed as it was stated. What was needed was the ability to see this not as a problem but as part of a situation that could be perceived in a variety of alternative ways. It required the courage to ask a different set of questions from those being asked at the time.[5]

Furthermore entire disciplines (and subdisciplines within them) are sometimes transformed not necessarily by new solutions to problems per se, but rather by new ways of "situating" (meaning that a problem/solution matrix is reconstructed as a situation) already solved problems in a new environment.

Such an orientation is mentioned in Brown (1996b) with regard to the Erlanger program in Germany in which Felix Klein in 1872 completely redefined geometry in terms of transformations:

> The Erlanger Programme launched in 1872 by Felix Klein in geometry was born not out of problem solving alone, but out of wondering how the field was being defined by its existing assumptions and array of problems. Even when we focus on "the field" and not the psyche of individuals, it is never problems and their solutions alone that enable us to classify the field of inquiry. Rather it is the human act of deciding what are pleasing, elegant, economical, challenging ways of viewing the field itself. We are never interested in solutions *per se*, but in solutions that are consistent with what researchers in the field think the field is about, and as Toulmin (1977) has shown, "aboutness" for fields has more to do with the questions asked of the field than with the objects themselves.(p. 1315)

Here again, we can construct a valuable pedagogical analogy—one that is an extension of what we discussed at the beginning of this chapter in the Jordan example of congruence for triangles, and in Debby's perception of what algebra was really all about. That is, it is not only interesting for diagnostic purposes to find out what a student thinks a unit or a field of inquiry is about. Rather, a class record of its perception regarding the nature of a topic being studied for several weeks, months, or even an entire semester would reveal not only how students see what it is all about in some general sense, but how their viewpoint relates to the problems they entertain.

Here it would be valuable to record dissenting opinions as well as popularly held points of view. It would also be enlightening to see how and why perceptions change over time. It is not that problem solving is irrelevant in defining and reconstructing what a field is all about. It is rather that a quite different *kind* of problem is addressed whenever an effort is made to make sense out of a variety of mathematical and educational experiences. We will be developing this idea a bit further in some of what follows, and in the sections entitled "Pseudo-History" and "On Error Making" in chapter 6. There we will provide a final gloss on the pedagogical problem of taking evolving conceptions of a field as a different sort of problem from the one depicted in Figure 4.1.

Revisiting Constraints as a Component of Problems

Having seen a variety of ways in which Figure 4.1 can be elaborated upon, even when solving is either downplayed or incorporated in a much more robust educational experience, we approach the issue once more by considering an important element of problem: the concept of constraint.

We spoke in the section entitled "Problem/Situation Reverberation" about the educational value of communicating with each other about what we see and what we notice. To talk about what is noticed or not noticed is to make salient constraints of a problem. Recall that Nickles defined a problem in such a way that it included all the constraints plus a demand for a solution. In doing so, he wanted to be sure that built into the problem itself were such matters as the history of the problem. From his point of view, to state a problem without its history is to admit that something about the problem has been left out. In fact, if its history is not built in even implicitly, then the alleged problem would not even be a problem.

We can, however, expand upon the concept of constraint considerably for educational purposes—regardless of how the matter is adjudicated in the field of philosophy of science. Doing so, not only has the potential to create a more humanistic environment, but also enables all of us to see the relevance of perspective in the world. As we have been arguing, nothing merely exists, but rather exists in some set of contexts.

What are some of these educational/humanistic contexts? As a start, let us reconsider the concept of history of the problem. Nickles and others who debate the issue from strictly a philosophical point of view have in mind such matters as where the problem came from and what headway had been made on it in the past. These issues reflect the importance of the history of ideas and its evolution, but they are interesting educational matters as well. Given a collection of infamous word problems (even some that are neither famous nor infamous, like the ski vacation problem in the section entitled "Problem/Situation Reverberation)" we can incorporate within the students' education the possibility of reflecting upon when problems like those arose in the curriculum, and what they were intended to accomplish. Might it say something about what educators believed was "relevant" to the real-world concern of youngsters? Could it be that some educator just happened to come upon an intriguing problem that she thought was a good application of something more abstract, regardless of its actual relevance to students' lives?

Though some of these questions may in fact have answers that can be verified by investigating appropriate sources (old textbooks, for example), we can also surmise reasonable answers that may in fact not be easily verifiable. Just as we suggested the value of imagining how a mathematical idea might have evolved in the section on scope and intentionality, here is a first-rate opportunity for youngsters and their teachers to discuss an issue about the nature of mathematics learning from possibly differing points of view. Since the participants know that the answer is not at all apparent to anyone, everyone has the opportunity of uncovering competing incipient theories of learning and teaching without necessarily bowing to an ultimate "expert." Furthermore, here is an opportunity to discuss how some hunches may be better than others, even though the actual issue is not easily resolved.

There are however, other analogies we can make with the concept of history if we view it from an educational lens. That is, part of the constraint on a problem is one's own history in encountering problems of any sort. What do I tend to do when confronted with problems of various sorts? Do I need to see some practical application in order to take it seriously? Do I need to see a larger picture—indicating where the problem came from and how it relates to a field? Do I have to visualize how the problem relates to other ones like it or do I try to see it "on its own" regardless of how it resembles other problems? Here we verge once more on the issue of perspective over time and of an evolving conception of a field that we discussed in the above section.

Furthermore, educators can encourage conversation about all of the above constraint issues not only with regard to problems but with situations as well. Thus we can discuss what appear to be the constraints of a situation such that we each see the situation from a different point of view. We could continue finding ways to expand the concept of constraint that is built into the definition of problem according to Nickles and Agre, but the realization that problems come with constraints and that people who encounter the constraints are part of the problem as well redirects what there is to investigate about anything that is supposedly a problem. Readers will surely come up with other interesting constraints, but what I am suggesting is that a conversation of constraints has the potential to open up options about personhood and about the nature of the discipline that we have not begun to imagine in the curriculum.

Though we have reached pedagogical suggestions in this section that are close to ones we reached in the previous two sections, it is worth uncovering nuances of difference and just as importantly, it is worth appreciating that they come from slightly different sources of concern. In one case ("Problem/Situation Reverberation"), we derived our suggestions from comparing what is *noticed* as problematic in given situations and problems. In another, we focused on an effort to analyze what is meant by progress in a field or topic ("Problem/Situation Reverberation and Progress in Mathematics"). Finally we took the concept of constraint that is built (sometimes in a subtle manner) into a problem, and we sought pedagogical analogs.

Different people starting from these three sets of concerns would very likely propose quite different educational programs. What I am suggesting here (and throughout the book) is not an appropriate set of educational conclusions, but rather a different set of lenses through which to view the venture. In an important sense, I am operating throughout in a self-referential way in that I am attempting to invoke the very concepts I am creating in the act of describing and elaborating upon them.

Tightening the Bond of Problem and Solution: An Ironic Twist

We take as our next point of departure a slight variation of the fallacy we identified in chapter 3.

Given any problem, in order for it to be used for educational purposes, it is necessary that the goal ultimately be directed towards (a possibly wide variety of) efforts of students to solve that problem.

A quite interesting and to some extent ironic variation of the model actually makes use of *both* a given problem and solutions associated with it. Instead of challenging the concept of problem "as given," we can take the problem/solution union as a point of departure. In what ways might we choose that pair as our unit to move beyond the narrow bond of problem and solution? What kinds of things could we invite students to consider if both a problem and its purported solution are given as part of the agenda? Much of good teaching, even if focused upon problem solving, does involve looking at how problems have been solved by others as a step in the process of encouraging independent problem solving. In our earlier discussion of how constructivism can go awry, we pointed out that people who listen to each other's solutions do not necessarily merely "absorb them." Nevertheless, to be given a problem and its solution would seem to undercut the alleged power of solving problems on one's own. What is left to do if a problem as well as the solution are given? Where is the motivation to solve a problem that has already been solved?

We explore these issues, not because we are claiming that some valuable thinking does not already take place in existing curriculum that begins with a problem and its solution as given, but because we would like to broaden the humanistic educational options that might flow from such a beginning.

We have pictured the set-up in Figure 4.5.

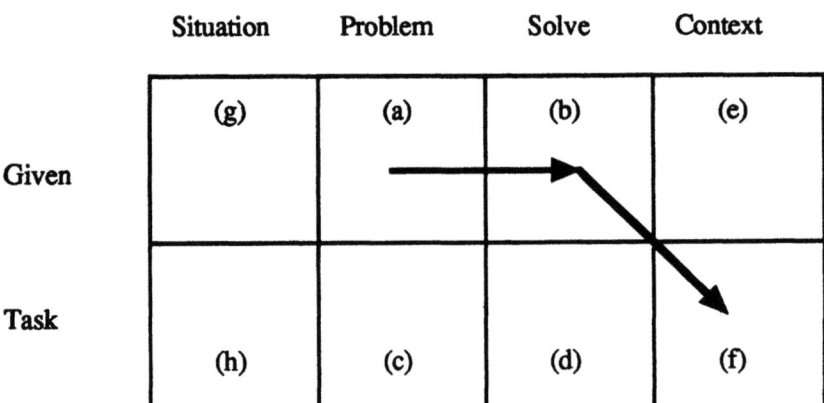

Figure 4.5 The place of context given both problem and solution.

Problem Posing/Solving in a Humanistic Light 99

What are we to make of box (f) as a task if we are given (a) and (b)? One possibility is that students could be encouraged to find other ways of solving the problem than the one presented. Alternatively, they could focus on *understanding* the solution—assuming that a solution does not necessarily make sense to them. But what other categories might be invoked if we adopt a humanistic perspective on this model? Let us look at one example, not for the purpose of creating a template but to seek some options.

One of the most interesting problems in the history of mathematics was solved by Euclid about two thousand years ago. It is a problem that has had ramifications in one form or other since, and it continues to raise some challenging issues regarding the nature of mathematical thinking, the growth of mathematics and in the relationship between mathematics and general culture.[6] Having achieved some understanding of what a prime number is and why it might be a significant topic to think about, many other problems come to mind. We might be interested in knowing how to generate prime numbers, how to determine if some (large) number is prime, how it is that prime numbers might relate to other areas of mathematics (e.g., geometry or calculus).

An intriguing question is: How many prime numbers are there? If the answer were: "a finite number," then it would be hard to imagine that prime numbers could be building blocks of natural numbers. Yet, surprisingly enough, inductive evidence appears to suggest that the number is in fact finite. Table 4.1 tells us how many prime numbers there are in different intervals for the first one thousand numbers:

In fact, it is easy to establish that it is possible to create as long a string of nonprimes in a row as anyone would wish.[7] Now let us look at Euclid's proof that there is an infinite number of primes.

It is a proof by contradiction. Euclid assumes that there is a finite number of primes and demonstrates that this assumption leads to a contradiction.

Table 4.1 The number of primes in several different ranges in N.

Range	Number of Primes
1–200	46
20–400	32
401–600	31
601–800	30
901–1000	29

The first few primes are 2, 3, 5, 7, 11, 13. Let us symbolize the first one as p1 (so p1 = 2), the second by p2 (so p_2 = 3), the third by p_3 (p_3 = ?), and so on. If we assume that there is a finite number of primes, then we can arrange them in ascending order until we reach the last one. Let these primes be denoted by $p_1, p_2, p_3, p_4, p_5, \ldots, p_n$ where p_n is the final prime.

Euclid's solution now suggests that we form a new number and look at it closely. The new number, K_n is formed by multiplying together the finite list of primes (assume for ease of following the calculation that n = 5) and then adding 1:

$$K_5 = (p_1 \cdot p_2 \cdot p_3 \cdot p_4 \cdot p_5) + 1.$$

Now let us look at this new number. It is surely bigger (why?) than any of the p_i's listed so far. What else, however, can we say about it?

K_5 is either prime or composite (i.e., a product of primes). Let us look at both possibilities:

i) If it is prime, it is bigger than any of the primes so far enumerated and we thus have a solution to the problem of finding a prime bigger than any of those listed.

ii) If it is not prime, then it must be a product of primes and hence divisible by *some* prime. This should not be hard to see intuitively. It is why we think of primes as building blocks of numbers in the first place. However, none of the primes listed so far will divide K_5 since each of the primes listed will leave a remainder of 1. In the case of p_1 for example:

$[(p_1 \cdot p_2 \cdot p_3 \cdot p_4 \cdot p_5) + 1] \div p_1 = (p_2 \cdot p_3 \cdot p_4 \cdot p_5) + 1/p_1.$

So there must be some other prime different from all those we have found so far.

This then is a modern day interpretation of Euclid's proof that there is an infinite number of primes. For the original, see Heath, 1956, Vol. 2, Proposition 20, p. 412. Well, so what? What does this have to do with a humanistic perspective? Nothing in itself. However, everything follows from how we think about what flows from this set-up. Again, though it is quite special, we use this problem and solution as a case study for how it is that the union of problem and solution still leaves much to do that does not fit the problem solving mode we have depicted in Figure 4.1.

Let us begin by observing that, though quite brief, this solution to the problem—"How many primes are there?"—is an act of genius. Students are not likely to come up with this solution on their own, though there is much for them to appreciate once they are presented with it. Above and beyond understanding the problem and the solution, what are some of the issues that might be discussed? Below are a few questions that are suggested by this solution (based upon some actual classroom activities):

1. There obviously is a difference between coming up with a solution and being able to appreciate it once it is presented. Why should it have been so hard to come up with the above solution when it is so brief and relatively easy to follow?
2. Are there circumstances in each of our (nonmathematical) lives that exemplify the above principle: something hard to come up with but relatively simple and easy to appreciate once we have been made aware of it?
3. Many people claim that this is one of the most elegant solutions to a mathematical problem. What qualities do you find here that might contribute to a feeling that this is elegant? Have you experienced the feeling of elegance in other aspects of your life? How does it compare with the feeling you (or others) have about the elegance here?
4. Frequently people experience mathematics as "plodding" and "obtuse." Have you come upon mathematical ideas other than this one that relate in an intuitive way to the concept of infinity. Might some of them fit the category of "elegance"? (Think about when you were very young and falling asleep . . . worrying about whether or not you might fall off the end of the universe.)
5. There is high irony in this solution from at least one point of view. In an effort to create this number K_5, Euclid first produced a number that could not be less prime. That is before adding the number 1 at the end, he created the number $(p_1 \cdot p_2 \cdot p_3 \cdot p_4 \cdot p_5)$. That number is the most nonprime number in Christendom.
   a) What might have been going on in Euclid's head that got him to make that move? Notice that he then added 1. Could he have added 2? 3?
   b) We are thinking of a temporal order of sorts in the creation of the number $(p_1 \cdot p_2 \cdot p_3 \cdot p_4 \cdot p_5) + 1$. Do you think that is a legitimate way to think of a number? More precisely, we can think of $(p_1 \cdot p_2 \cdot p_3 \cdot p_4 \cdot p_5)$ as being the first move and adding 1 to be the second move. From what points of view does this temporal concept make sense? What alternative ways are there of thinking of the number K_5? (again realizing that 5 could be replaced by any value.)
6. This solution to the problem is a bit slippery. Many people think that they understand it one minute and lose it the next? Have you experienced this reaction in this case, or in others? Why do you think some people have the feeling of slipperiness here? What are some things anyone who felt that way might do?

7. A lot of people who see this proof for the first time think that the argument is that either i) is the case or ii) is the case. They believe that Euclid is claiming that ii) cannot be possible, and therefore we are left with i)—that K_5 in fact must be prime. Why do you believe people have this conception? Did you have it?
8. Having the above conception suggests new problems to pose that are not explicitly mentioned in this solution. One possible problem would be: Is it possible that the conclusion in 7 is in fact a correct conclusion—even though Euclid's proof may not be demonstrating it? How would you find out if this conclusion is justified independently of Euclid's proof?
9. Before Euclid came up with the proof that there exists an infinite number of primes, you can imagine that some people (maybe even Euclid) were in doubt about the answer to the question "How many primes are there"? Can you put yourself in a position of such doubt?
    a) How would you create such doubt for yourself?
    b) Suppose there in fact were a finite number of prime numbers. What might the consequences be, and consequences for what? Where would you look in attempting to find an answer to the question?
10. We have introduced the concepts of
    a) elegance,
    b) "slipperiness,"
    c) "psychoanalysis" of the author's thoughts in coming up with the proof,
    d) doubt that the problem has the solution it does, and
    e) the irony in this example.

    How do you react to these categories being imposed on mathematical thinking? Do you think these categories belong to the mathematics curriculum or are they perhaps beside the point? How do they affect your ability to appreciate mathematics and to appreciate your own mind and feelings?

Though some of the directions summarized in (10) may enhance our ability to understand a bit of mathematics, that is not our *only* educational goal. We are not suggesting that the major reason for exploring question (10) above, for example, is to enable students to better appreciate Euclid's proof of the infinitude of primes. It is one possible consequence, but there are others.

While issues of problem solving and their heuristics and of understanding are included in the issues generated by the above questions, it is clear that much more is going on. It is worth observing that the perspectives introduced are broadening the standard (and also *The Standards*) terrain n-fold.

We are inviting students to imagine what might have been in other people's minds when they were forming their ideas; we are inviting them to react to the intellectual and emotional dimensions of their own experiences in relation to the mathematical concepts that are being developed; we are asking them to use what many people would dismiss as errors for the purpose of expanding the kinds of questions and analyses they might consider; we are asking them to appreciate that ideas come in bottles of different sizes and shapes, and that some are prettier than others and some are more inspiring than others. Most importantly, we are inviting them to use this mathematical experience as a stepping-stone to consider important related issues in their own lives and in their culture—issues that may be inspired originally by a focus on mathematical experiences, but that transcend that focus in favor of personal and educational issues.

Revisiting Problem and Solution as *Given*

We have not so much eliminated the connection between problems and solutions as we have suggested loosening of the bond between a given problem and its solution. In the above Euclidean example, we have chosen the problem (determining whether or not the number of primes is infinite) and one purported solution for the purpose of defining a new set of problems, some of which relate to the mathematical analysis and some of which outstrip that analysis.[8] Thus, if we call the original problem, *problem*$_1$ and the solution of Euclid's, *solution*$_1$, then we are suggesting that an important educational agenda is served by taking the union of *problem*$_1$ and *solution*$_1$ as a new starting point. But starting point for what? We can think of the union as representing a *situation* which itself can then inspire the statement of some new problem of a mathematical or educational/personal agenda.

While we certainly know that knowledge grows by creating new problems based upon the solution of old ones, we have been reluctant to find within the problem/solution resolution the potential to create problems that outstrip a parochial sense of growth—one that essentially sees a solution as implying a rather narrow set of new mathematical problems.

With our eyes glued to a technocratic rather than a humanistic perspective, we not only limit the educational problems we are inspired to

$$[\text{problem}_1 \text{ and solution}_1] \longrightarrow \text{situation}_1 \longrightarrow \text{problem}_2$$

Figure 4.6 Joining problem and solution to create situations and problems.

explore, but we see the mathematical ones in a narrow light as well. We summarize the above remarks in Figure 4.6, where *problem$_2$* may or may not be a mathematical problem *per se,* but it was inspired by the entry before its arrow.

It would be worthwhile for the reader to reexamine the ten categories we considered following Euclid's proof in order to explore the sense in which the new explorations derived from the problem and solution were mathematical issues, issues dealing with mathematics but tangentially related to the analysis, or issues that were educational but not at all mathematical. Furthermore, it would be valuable to reexamine these ten categories not only for the purpose of determining degree of mathematical relevance, but also for the purpose of teasing out to what extent the questions are content specific, a function of the specific problem we analyzed, and to what extent they can be generalized.

Rather than engaging in that reexamination at this point, we turn instead to a consideration of ways of thinking that do have considerable potential for raising humanistic questions based upon reflection upon the "givenness" of the problem/solution interaction, i.e., creating situations, that are not normally invoked in the mathematics curriculum.

We turn first to the category of wonder, a category that speaks to the issue of mind and emotion. We then will examine the concept of "same/different," a dimension that appears to be more cognitive than wonder. It bears on the ways we compare new and emerging ideas. Both will exemplify what we have called humanistic educational dimensions.

Problem, Solution, and Wonder

One of the most repressed yet fascinating educational contexts that might be selected in the scheme of Figure 4.5 is that of wonder. Wonder appears to be yet one more instance of what Scheffler (1991) refers to as a "cognitive emotion." He has looked at the joy of verification and the feeling of surprise as examples that belong to this category. At first blush, calling something a cognitive emotion may sound like an oxymoron. In referring to an emotion as cognitive, however, he is not trying to reduce the emotion to a cognition, but rather claims "that . . . it rests upon a supposition of a cognitive sort—that is to say, a supposition relating to the content of the subject's cognitions . . . and, in cases of special interest . . . bearing upon their epistemological status" (p. 9). Though underrepresented in curriculum, there is much to wonder at in any field and mathematical thought/experience is no exception.

In order to appreciate the mathematical sources of wonder, it is necessary to think a little about the relationship between wonder and knowledge. Green (1971) addresses this issue with eloquence. His philosophical discussion begins with some ordinary language analysis. He notices that wonder is an expression that does not come unqualified. Looking at the linguistic constructions of wonder *how* or wonder *why*, he comments, "Curiosity, the capacity to wonder how or why, has its roots in a kind of ignorance. It stems from our recognition that there is some feature of the world we do not understand but, given time, can comprehend. The capacity to wonder *how* or *why* stems from the fact that our knowledge is incomplete" (p. 198, emphasis added).

So, knowledge quenches our inclination to wonder *how* or *why* something is the case. There are times, however, when our wonder is not quenched by knowledge, and we continue to wonder even after we in some sense have an explanation or a supposed solution to a problem. This is the case when we wonder *at* (vs. *how* or *why*) something. It is essentially an admission of amazement. It is not because we *lack* knowledge, but rather because we *possess* knowledge of a certain sort. What is that sort? Let us hear again what Green says about this form of wonder and then consider the special sort of knowledge one has to have in order to be amazed. He says of wonder *at*:

> An advance in knowledge may remove . . . a disposition to wonder. But it will do so in a different way than if the roots of wonder lay wholly in a kind of ignorance. With greater understanding I may come to comprehend some explanation as to why another should love me and I should love an other. But for all my knowledge, love will remain a marvel. I may be disposed by learning to forget the fact, but other things than learning may have the same effect. The wonder of it may become hidden from me. But this means simply that, though wonders may cease, they do not cease because we know more about them. They cease because we become blind to them. To wonder in this sense is not to be curious—to wonder why or how. It is to wonder at or to wonder that. It is not primarily to ask a question but to marvel or to be amazed. That is why one may wonder *at* long after one has ceased to wonder *why* or *how*. (p. 196, emphasis in original)

What are some things about which one may wonder *at* in mathematics even after explanations have been provided? One of the most interesting sources of wonder for me is the connection between two ostensibly distinct branches of calculus—integration and differentiation. One branch can be thought of, in a concrete sort of way, as being concerned with finding out something about the inclination that a straight line makes with a horizontal axis when that line is tangent to a curve at a point. Another branch is concerned, again as a concrete realization, with finding the area under a curve bounded in a certain way. Two quick sketches of the two branches are depicted in the sketches of Figure 4.7 and 4.8, respectively.

Figure 4.7 depicts differential calculus. The angle ß is the inclination that the line μ makes with the horizontal if μ is tangent to a point, A, on the curve. Figure 4.8 depicts an area (the shaded region) under the same curve. Though both concepts are geometric and deal with limits, they appear to be unrelated in other essential respects. What would the inclination of a tangent line to a curve at a point have to do with an area under a curve? One is a linear measure and the other is a square measure.

Yet the two concepts are in fact related by what is known as the Fundamental Theorem of Integral Calculus. The exact statement of the relationship is unimportant. What is important is that two concepts that appear to bear almost no relationship to each other are integrally (pun intended) connected.

It is possible to view the statement of their connection as a problem solving activity. There are many ways of showing the connection, but the

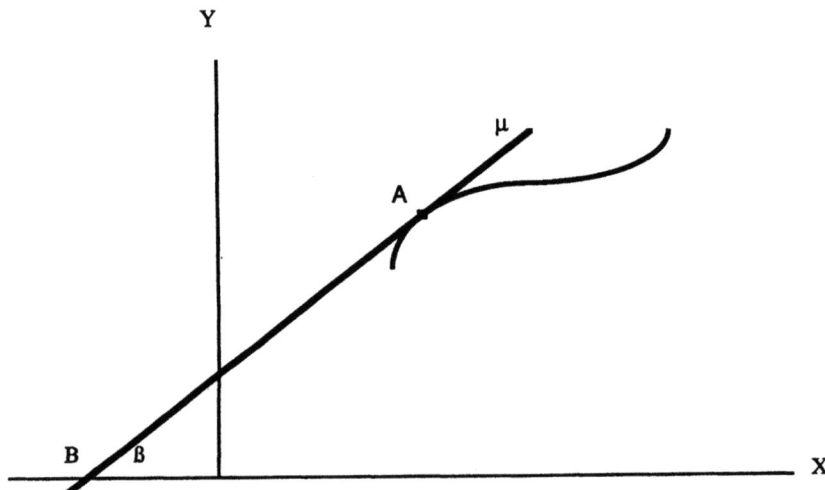

Figure 4.7 Depicting the derivative as the slope of a tangent line to a curve.

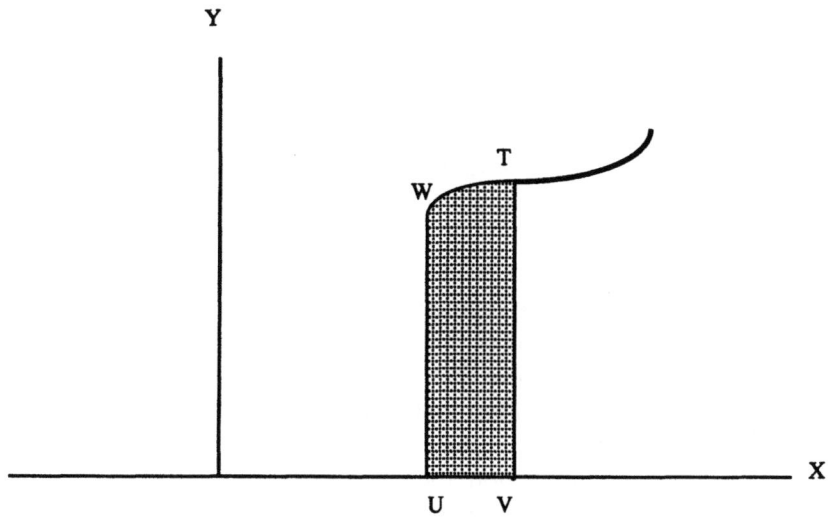

Figure 4.8 Depicting the definite integral as the area under a curve.

point is not so much that we can solve the problem of finding the connection, but rather that no matter how we solve it, we continue to be (at least I am) amazed at the realization that they are so related.

Why am I amazed by the connection even after I have engaged in a problem solving act that states precisely and even proves the nature of the connection? Explanations and solutions help, but no matter how many there are, I am still haunted by the earlier intuition that the ideas are independent of each other. Proofs and problem solving activities are one thing. Expectations and intuitions are another. There are surely occasions when proofs and problem solving transform our intuitions so that we no longer hold onto what may have been naive expectations. That is part of what we mean when we claim that someone has been educated.

It is equally powerful, however, to be able to remember worldviews we held before careful logical arguments enlightened us and demolished them. They are part of who we were. We are in a constant state of transforming our large and small intuitions about the world. What is needed in order to maintain a sense of wonder, however, is a protection against amnesia. In many ways being educated not only transforms the way we see things but involves a loss of ability to be our former selves. As hard as we try, for example, to read a sentence as if it is nonsense after we have been educated to read, we cannot do so. Part of being a person is not only growing to see things differently but seeing with some degree of clarity where it is we were, and how and why we moved to a different position or different worldview.

To be amazed even after the act of problem solving frequently requires that we allow two worldviews to reverberate rather than to have one subsume the other. Every time our intuition is educated by some rigorous problem solving act, we can profit from attempting to see how the two worlds relate and even collide.

Examples with infinity abound as illustrations of this sort of collision. Look at Figure 4.9. It is a line segment two units long, from 0 to 2. There are many points along the line segment that can be depicted by fractions. It is not hard to come to believe that there is an infinite number of them, though surely that is an intuition that some people may lack. Now suppose we wonder about the existence of numbers in that interval that cannot be expressed as a fraction, the quotient of two natural numbers. It is not so easy to come up with such numbers.

It is not difficult to persuade yourself that there appears to be a number somewhere in the interval which when multiplied by itself yields 2. Yet it is a highly frustrating task to figure out what precise fraction will do the job. [We might try a number like 1.4, which as a fraction would be 14/10. If you take 14/10 and multiply it by itself, you do arrive at 196/100, which is almost 2, but not quite. If you take 3/2 and multiply it by itself you get 9/4 = 2.25, which is bigger than the desired number]. Try to find something then between 1.4 and 1.5 that might yield 2 precisely when it is multiplied by itself].

After a significant amount of empirical investigation, a proof that there is no such fraction (i.e., no rational number) is both revealing and unsettling. It is a first step in coming to appreciate that there may be some numbers that cannot be expressed in the form of a fraction, or decimal, at all.

Now if we wonder how many such numbers (ones that cannot be expressed as fractions) there are in that interval, we surely would guess that the number is limited. Anyone who thinks about this for the first time would surely conclude that the number is limited and would probably have much difficulty finding even one such number. Well, it turns out that not only is the number of such numbers unlimited, but in some sense there

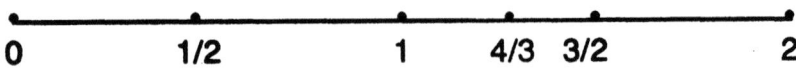

Figure 4.9 Numbers between 0 and 2 as points on a line.

are more of them in that tiny interval than there are all the fractions in the universe!

All of this may sound cryptic to anyone who has not explored problems of this sort. Alleged proofs that such weird numbers exist may appear to be an act of legerdemain when first introduced—proofs such as the ones that demonstrate that the square root of 2 is irrational and that the number of irrationals is uncountable.[9] There are, however, fanciful ways of introducing these ideas (even in the absence of solving problems) that enable us to appreciate what the intuitions are that are in need of reeducation. One way of seeing what is meant by claiming that a number cannot be expressed as a fraction follows.

Suppose you have an apple orchard with trees equally spaced as depicted by the dots in the array of Figure 4.10. Assume that the dots are very small, however, so that the trees have position but no thickness. Suppose you are situated at point A and look out at an infinite expanse of such dots. If every line of vision from A into the orchard is depicted by a straight line, can you imagine that there is some line of vision that has the property that you would never hit a tree no matter how far you looked? A

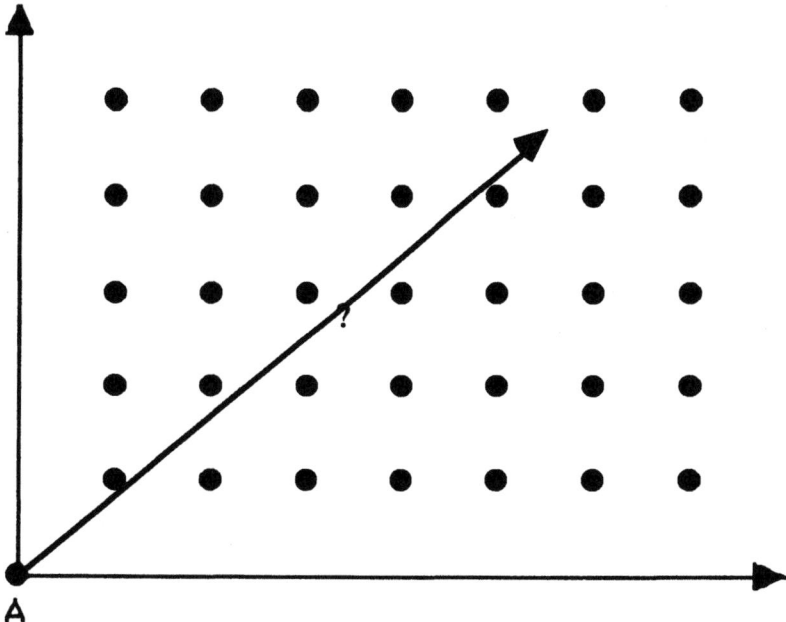

Figure 4.10 Depicting lines of vision from the origin in an apple orchard.

lot of experimentation would lead you to believe that no matter what line you chose as your line of vision, that line would eventually hit a tree (something that certainly is the case for many lines from the origin).

The claim we made above, however, with regard to the irrationality of the square root of 2 conveys the opposite. Not only is there one line of vision from A that would never intersect a tree no matter how far out we looked; not only are there an infinite number of such lines, but in fact there are more lines that will never intersect a tree than those that will.[10]

Some indication that we may profit from wondering *at* the concepts that have been revealed over the centuries (with regard to existence of numbers beyond the natural numbers of our youth) can be found in the names that have been given to such number systems. We have been discussing the case of "irrational" numbers.

Such language (to be discussed further in chapter 5 on metaphors) is a clue that the history of number systems has been filled with drama and controversy. In every case, people found it helpful to work with newly emerging conceptions of number but at the same time were reluctant to admit that these new quasi-giants/quasi-beasts were in fact numbers.[11]

One of the most interesting historical regarding how it was that competing conceptions of extended number systems were adjudicated was the realization that it often was not greater rigor that won the day. Rather, as Rogers (1964) has shown, frequently the conception of an extended number system that won favor was the one that accorded best with the intuition of the times rather than the one that was most rigorously established by solving problems.

To Wonder Where We Have Been Headed

We have not so much eliminated an interest in solution as it relates to problem, but rather we have taken a problem and seen a solution to the problem not as an end in itself, but rather as a source for further humanistic inquiry. By juxtaposing the solution against our early intuitions about the solution and by keeping the two in mind simultaneously, we come to appreciate many things. First of all, we become more sensitive to our intuitions themselves. Second, we put ourselves in the position not only of realizing that our intuitions frequently are challenged by more rigorous conceptions, but we need to decide when and if we ought to relinquish or modify these intuitions in the light of such rigor. Most importantly, we place ourselves in a different epistemological/psychological state by so attending to what we have called the reverberation between early intuition and later solution.

By framing the issue in the context of wonder, we realize that an increase in knowledge does not automatically quench our thirst for inquiry. No matter how much we know, we can find ways of seeing knowledge, not as an end in itself but rather as a source of self-awareness and self-understanding. Wonder raises the question of where we have been and where we are headed in relation to evolving concepts. It is a phenomenon that encourages us to see the present and the past as a cloth.

When we select issues connected with infinity in exploring the potential of wonder in relation to problem and solution, we have tended towards the spectacular. Green (1971), however, points out that we make a big mistake when we focus on the extraordinary as a source of wonder. He says,

> We suppose for several reasons that wonder . . . is aroused only by spectacular or extraordinary things. But that is a mistake. We may wonder at travel in outer space or marvel at the prospects of a journey to the moon. But shall we continue to wonder at the hundredth flight around the earth or the thousandth picture of the moon's far side? Wonder aroused only by sensational things is satiable, because they have a disgusting way of becoming usual and ordinary. When men find occasion to wonder only at the extraordinary or spectacular, it is the surest sign that wonder is already dead. (p. 196)

What is it then for Green that is the source of wonder in the ordinary? It has less to do with the regularity or rarity of its occurrence than with the fact that it happens at all. He comments,

> The wonder that is ceaseless, that can never be exhausted, has always to do with what is usual and close at hand; for the marvel of a thing has less to do with its frequency than with its contingency. It is both remarkable and ordinary that we plant an apple seed we get an apple tree. We may think there is some necessity about the sequence we all observe: that apple seeds produce apple trees which produce apples, followed by small boys and small wars and apple cores. But there is no necessity in this sequence at all. We can imagine an apple tree that produces pears or peaches or, for that matter, magic apples or magic pears. That is the charm of what Chesterton has called the "Ethics of Elfland." (p. 197)

While having drawn our attention to the ordinary as a source of wonder, he has overstated contingency—that things need not be as they are—as a criterion for wonder. In asking, for example, what it is that may not be a source of wonder, he selects logical necessity as an unchallengeable domain. He says, "What then may we take for granted? Nothing, except those logical necessities which are stringently observed even in fairy tales" (p. 197). I suspect that he includes mathematical thinking when he speaks of logical necessity, even though logical necessity may be a bit of a fiction within mathematics. Assuming however that the concept has validity in mathematics, is there reason to depict what is logically necessary as an invalid source of wonder?

In our analyses above, we have shown how the interplay of intuition and rigor can be a powerful source of wonder. If we have chosen the spectacular for that source, wherein fits the ordinary? A return to What-If-Not sort of thinking reenergizes the concept of wonder in the ordinary. That is the smallest of changes in anything that is "given" as ordinary and rather dull can generate spectacular differences. By "slightly" changing the domain of prime numbers we have seen that some of the most humdrum expectations receive miraculous rejuvenation. In chapter 2 we showed that unsolved problems, like the search for a generating formula for prime for all prime numbers in the set of natural numbers, curl up in embarrassment in set E.

The realization that a connection is not a necessary one in the real world becomes a source of wonder for Green. However, it is no less a source of wondering *at* to appreciate that the slightest modification of any phenomenon raises the possibility of opening up entirely new expectations and unimagined territory to explore.

In this regard it is interesting that Henderson (1996) in a provocative essay, asks the rather innocent sounding question, "How could I, an expert in geometry, learn from students"? He concludes that it is by carefully listening to them as they try to offer proofs based upon assumptions and even worldviews that are different from his own that he, an expert in geometry and veteran teacher, comes to new understandings about the meanings and interpretations of what is proven to be so.

What he does not acknowledge, however, is that even, perhaps especially, relatively naive students are in a position not only to come up with unusual proofs, but to pose spectacularly new problems to explore if they are given the encouragement to do so. Furthermore, while experts may have the competence to solve many of these problems in the sense that novices do not, they may not necessarily have the special ability to *generate* these problems. In fact the familiarity of experts may inhibit their ability to generate what only the novice can question.

Perhaps the greatest source of wonder resides in the realization, if not the expectation, that our relatively naive students may enable us to see interest and significance in what we never had seen before. "Seeing" begins with *noticing*, and much of what we do not notice (anymore) is often a function of the excess baggage of knowing a great deal.

The Concept of Same and Different

One of the most interesting ways of thinking about problems is to perceive of them as objects to be *compared* rather than as *demands* to be *fulfilled*. To compare problems, however, requires that more than one problem be an object of inquiry. There are many ways of comparing problems. Some may be more complicated than others; some may be more appealing than others; some may be longer than others; some may be more transparent than others; still others may (according to "relevant" criteria) be more relevant than others.

As in the section on wonder, we are not so much interested in eliminating a focus on solutions, but rather, as depicted in Figure 4.6, in looking at problems in a light that transcends their solution per se as a focal point or as a final arbiter of what to believe. We are seeking ways of creating new problems and situations from old ones that allow for a variety of insights that outstrip the normal sense of "progress" associated with mathematical inquiry.

We set the stage for comparison of problems by relating the following story:

A Story of Same and Different

Many years ago, before qualitative research in education had gained the respectability it now enjoys, I attended a conference on research in creativity. Most of the reports were quantitative studies designed to determine the role of creativity as a dependent or independent variable in relation to some teaching regimen. One presentation stood out, not only because it was different in design, but because of the speaker's scope, courage, and insight. Rather than reporting on a study, he told a story about the nature of his own research.

He was an educational psychologist, and he told us that he had written a number of essays in which he compared the theoretical orientations of different researchers. After a number of years, he came to realize that there was a mind-set that had pervaded all of these essays. What was it? He prefaced the personal realization with a characterization of the work of other people who were engaged in similar analysis. That is, most of them sought ways of locating subtle *differences* between researchers who others had seen as quite *similar*. He, however, found himself operating in the other direction. He sought areas of similarity that were viewed by other researchers as quite different. Seeking what he considered to be fundamental similarities between Skinner and Freud is a case in point.

In preparation for the conference on creativity, he had decided to try to figure out what accounted for this apparent professional anomaly. He did so not by trying to locate some respectable psychological theoretical perspective that would justify his stance, but rather by searching for personal themes that had affected his life. He reflected on his childhood, and tried to figure out what took place in his upbringing that might explain such behavior. Thus, he recalled that when he was two years old, his parents were divorced, and he was brought up by an unconventional pair of adults: his mother and his paternal grandfather. Apparently they never got along with each other. From the time he was able to recollect, the atmosphere of the household was explosively charged. His mother would come to him to complain about his grandfather. She would tell him, for example, how the grandfather wasted so much time by playing the phonograph.

Conversely, his grandfather would come to him and complain about his mother. He would tell the boy that his mother always dawdled in preparing meals, probably to get the grandfather's goat when he was hungry. What the speaker recalled himself doing as a young boy was trying to make peace. How did he do it? He found himself telling his mother that the grandfather was really not so different from her. Even at a young age, he tried to seek some "higher ground" that would demonstrate how the grandfather really behaved in a way that was similar to that of the mother. Similarly, he found himself telling his grandfather that his mother had much in common with his grandfather.

While the motivation for seeking similarity at such a young age was essentially a pacifistic attempt, he came to the realization that his professional orientation—the desire to seek essential similarity between apparent differences—was a generalization of that way of behaving, though the motivation may have changed.

Regardless of the veracity of the hunch, it was a fascinating one. It resulted in serious discussion that piqued the audiences' interest far more than any of the other presentations. It set me thinking in a different way about the concept of similarity and difference, and it is this evolving interest that I would like to explore in this section.

Whatever the criteria however, "to compare" means—at least implicitly—to ask how objects or ideas are *similar* or *different* from each other. There is perhaps no other quality that better defines us as human beings than the ways we handle similarity and difference.

Same/Different and Human Nature

The concept of same versus different seems to be built into nature itself and most likely has a direct impact on the evolution of a species. One of most instinctive reactions of each organism is to determine whether a new and heretofore unrecognized creature is friend or foe. The well known fight or flight reaction is predicated on the assumption that an organism is being threatened by another creature because it is different from itself in some important way.

Some organisms have a built-in imprinting response. Thus, ducks not only will follow their mother wherever she goes shortly after they are born, but they also will do so with other "organisms" that are similar in important ways to the mother. They even will be compelled to follow inanimate objects that bear similarity to mother ducks. Of course, the critical question here is: What means "similar?" It turns out that in the case of ducks what attracts them to a mother "look-alike" is not shape or size, as we might predict. Research studies have shown that it is the similarity of the "organism's" pitch to that of the mother duck that determines whether such imprinting will occur. Thus a baby duck will imprint a roboticized tape recorder if the sound on the tape is the same pitch as the mother's.

Cultures have fine-grained sensitivity to the concept of similarity and difference. Many outsiders would be struck by the similarity in looks and culture between Arabs and Jews; Turks and Greeks; North and South Koreans; citizens of Northern and Southern Ireland. In-group members in each of these cases, however, are acutely aware of the differences. They are so acutely conscious of them that people marshal all sorts of means to intimidate, harass, and even to eradicate others by virtue of the fact that they are members of the out-group. (Recall the Emo Phillips quote that begins this chapter).

Part of what it means to be tolerant is to perceive these differences and not to have them matter in some important ways. Thus, it is a significant first step for members of an in-group to realize that others may be different but not to be discounted by virtue of the difference. The concept of equal protection under the law is to acknowledge that there may be major differences among people, but by virtue of the fact that they are human, they deserve equal protection.

Part of the role of education in fact may be to discourage us from perceiving *some* differences all together. Of course though, we would not survive very long if no differences were perceived. It is important for me to be able to tell the difference between my wife and someone else's wife from some very important points of view. It is important for a mother to be able to identify her own child and to decide to nurse it rather than to select her grandfather for that purpose, for example.

Furthermore, no one would ever be able to read if all letters of the alphabet were considered to be essentially the same. To be an expert in any field means, among other things, that we can discriminate and see differences that others do not see. One x-ray of the heart is as good as another to the nonspecialist. Likewise, some experts see similarities among entire domains that others perceive as different. An expert in behavioristic psychology, for example, is struck by the similarity in behavior of humans and pigeons from the perspective of the role of reinforcement.

Thinking about similarity and difference is at the heart of what it means to be both human and humane. Granted that mere talk is not enough, how do we learn to talk about the concepts of similarity and difference in such a way that the talk affects the ways in which we think and act for the better? What does it mean to say that some things are the same and different in the first place?

In the next three sections, we will suggest how it is that the concept of "same/different" is deeply embedded in much of mathematical thought, though it tends not to receive the fanfare it deserves when the focus is more on problems and their solutions than on comparisons of problems and of solutions. While exploration of some of the edifices of the sort we are about to describe was an essential feature of the new math curriculum of the 1950s and 1960s, it is less honored but still present in *The Standards*.[12] Nevertheless, it is the bread and butter of much of collegiate mathematics and will most likely be resurrected again in secondary school education once it is appreciated that the study of system and structure need not be "imposed" but can be introduced with minimal use of technical language, with greater playfulness, and with more attention to selection of problems that are politically and socially appealing to a wider audience.

In the section entitled "Unearthing Implicit Ingredients of Comparison," we move to terrain that has received minimal attention in any of the reform movements. We select as a case study the notion of similarity and difference among three mathematical problems. In so doing, we will not only come to the realization that similarity exists where it did not appear to do so earlier, but more importantly, we will suggest how we might talk about the concept of similarity so that it has not only mathematical but personal import. This is not meant to be a lesson in pedagogy per se. No effort is made to suggest how the fundamental ideas might play themselves out in a specific classroom or text. It is intended, however, to lay out some interesting terrain within which problems can be used in some significant way that may incorporate solutions but that transcends a narrow focus on "getting a solution." In Appendix B, we show how the concept of "sameness" can be incorporated in an unusual textual manner, the format of the Talmud.

On Congruence and Equivalence Relations

The concept of congruence for geometric figures is an early historical mathematical example of what has become a hall-mark of mathematical thinking. Though Euclid thought of establishing congruence of, e.g., triangles by actually superimposing one on top of the other, he considered the idea so intuitively obvious that it never became an explicit part of his axiom system. A modern-day treatment differs from Euclid's perspective in that it sets forth the geometric transformations that can be used (rotations, reflections, translations) in order to place the intuitive concept of superposition on a firm ground. Despite these different points of view, they share a common assumption: They are both trying to locate the sense in which two *different* geometric figures (located in space in different positions) are "essentially the same."

The same concept is exemplified not only in geometry but in number theory as well—in the case of modular arithmetic. Consider, for example, breaking up the set of numbers into classes as follows:

A = {0, 3, 6, 9, 12, 15, . . .}
B = {1, 4, 7, 10, 13, 16, . . .}
C = { 2, 5, 8, 11, 14, 17, . . .}

An intuitive way of thinking about this partitioning is to imagine that at first, we have a simple clock consisting of three hours 0, 1, and 2. With our standard clock, we begin with 12 and rotate clockwise to 1, 2, . . . 11, 12 and then back to 1 again. The modified three-hour clock is depicted in Figure 4.11. below.

The twenty-four-hour clock continues in the same spirit as the 12-hour clock, but distinguishes night from day by depicting afternoon and

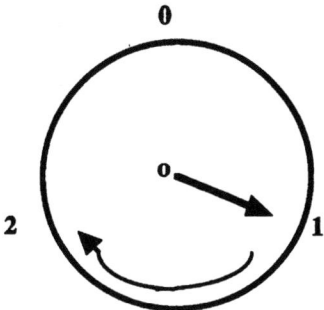

Figure 4.11 A clock with numbers 0, 1, and 2.

evening hours with numbers 13 through 24 o'clock. This method suggests how we might wish to extend the three-hour clock by associating 0 with 3, 1 with 4, 2 with 5, and so on as we continue to circle in a clockwise direction. Thus we would have the clock depicted in Figure 4.12.

The array of numbers associated with any number in that array (say 2) are all different. So, the number 2 is not the same as 5, 8, or 11. The numbers are all different, but they are the same from *some* point(s) of view. What is that point of view? There are many ways of expressing the sense in which they are the same despite their difference, and what you decide to choose as the "essential similarity" and how you express that similarity will depend in part on the amount of technical language you have acquired and also upon your intentions.

One interesting observation is that they are all the same with regard to their remainder when divided by 3 (for in a sense, divisibility by three is intuitively captured by the wrapping-around-the-clock way in which we created these sets in the first place). More precisely, the numbers in any given array are all different numbers, but they have the same remainder

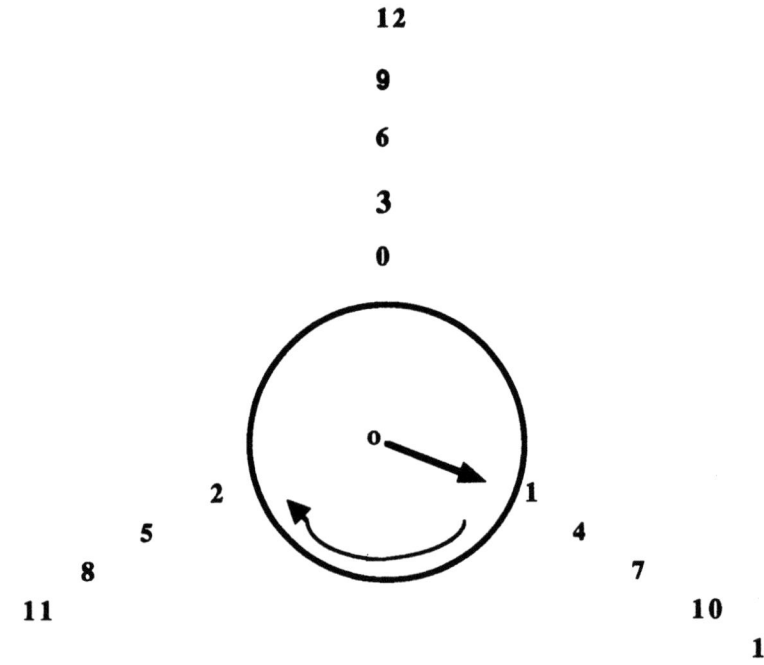

Figure 4.12 Wrapping all natural numbers around the clock of Figure 4.11.

when divided by 3. An interesting consequence of this observation is that all of the numbers in this array are distinguished from the numbers in the other two arrays with regard to the property of divisibility by 3.

Such a humble beginning has unusual consequences. As with the case of congruence for geometric figures, we have a form of congruence for number systems. In addition, we have the beginnings of a system that not only displays essential similarity among differences, but, in addition, we have an *operation* on the system (addition "of sorts") that enables us to join elements together in an interesting way. That is, 1 is not only like 4, and 5 is not only like 11, but 1 + 5 (i.e., 6) is very much like 4 + 11 (i.e., 15). How are they alike? The two different results (6 and 15) also belong to the same array. Furthermore, this phenomenon generalizes in a way we now depict.[13]

The "of sorts" in the above paragraph allows us to define a *new* operation of "addition" whose elements are not individual numbers but infinite sets—three of them. Any one of the three sets of an infinite number of elements can be thought of as generated by any of the elements of the sets. Thus, the fact that the equivalence relation is uniquely defined with respect to standard addition, +, enables us to define a new operation of addition on sets. So if [a] stands for the set of all elements that belong to the same class as a, and [b] stands for the set of all elements that belong to the same class as b, we can define

$$[a] \oplus [b] \text{ as } [a + b],$$

where the new operation $\oplus$ is defined (on sets with an infinite number of elements) in terms of the old operation, +, which was defined for pairs of numbers in N.

In this somewhat long-winded digression, we have captured a number of ways in which the concept of same/different functions. The first is essentially a classification scheme. We can partition the set of integers into three infinite sets by invoking the property of "remainder when divided by 3." Thus all numbers that have the same remainder are considered to be "essentially the same" despite the fact that they are obviously different numbers.

Even so, we have gone one step further. By creating a new system of the three elements, each of which is a set consisting of an infinite number of elements as in Figure 4.12, we have devised a new system with a new operation.[14] It is this kind of activity that is at the heart of extending number systems from ones that are known to ones that are on the verge of being created (for example from the set of whole numbers {0, 1, 2, 3, . . .} to the set of integers {. . . −3, −2, −1, 0, +1, +2, +3, . . .}).

Here then we have another essential concept of same/different that is a part of the mathematical backbone. That is, we have a relation (called an equivalence relation) that works in such a way that these different elements in the same class act like indistinguishable clones from *some point of view* with regard to an operation, in this case standard addition. It is much easier to understand what is being asserted once we can use stronger symbolism, but we have tried to treat these notions of "same/different" in a more chatty tone.

One way of summarizing what we have done in Figure 4.12 with regard to the new operation of $\oplus$ is via the "addition" in Table 4.2 (where [0] = {0, 3, 6, 9, . . .} ; [1] = {1, 4, 7, 10, . . .} ; [2] = { 2, 5, 8, 11, ...})

The richness of Table 4.2 is exposed as we investigate its properties. One obvious property is that we have a number that acts like 0 in the set of integers. That is, there is a number (which one?) that has the property that it is powerless in the sense that when added to any element, the sum is the same as it was before it did its thing. Another property that this simple system enjoys is an "anti-MacBeth" property in that it challenges the notion that "what is done is done, and cannot be undone." In other words, for every element in the system, we can find another element to add to it so that the result is to "zap" the original element and create a state of powerlessness.

So what? While it is beyond the scope of this work to further explore the significance of this system, we should mention that it is a simple example of a mathematically powerful structure, a group. From our point of view, however, it is enough to use this example as an entrée of sorts. It captures an essential quality of mathematical thought—the location of deep similarity in the midst of ostensible difference. We might wish to return to a variation of Figure 4.6, which depicts how problem solving can be used to create a new phenomenon that is on a different plane from the original problem and solution. After the dust is cleared, the new ob-

Table 4.2 Addition table for the three sets of Figure 4.12

$\oplus$	[0]	[1]	[2]
[0]	[0]	[1]	[2]
[1]	[1]	[2]	[0]
[2]	[2]	[0]	[1]

ject created by working through how clock arithmetic relates not only to three, but to all the natural numbers, is not so much of a problem but a situation. This situation is on a different plane from that of our starting point in that we now no longer solve anything but rather observe the concept of same/different as an element of thinking in mathematical and nonmathematical contexts. Now we turn to yet one more "same/different" category that is an inherent quality of mathematical thought—the concept of isomorphism. After having reviewed the following two sections, it might be worthwhile to return to the present one. Though we have not elaborated on the relationship between clock arithmetic and the system we have formalized in Table 4.2, the two systems are themselves an example of isomorphic structures.

On Isomorphic Structures: A Well Known Choice

Take any equilateral triangle and place a pin in its center as indicated in Figure 4.13. If you "spin" the triangle 120 degrees clockwise around the pin, you end up with the left-hand triangle in Figure 4.14 (so that A is now where B was before).

If you spin it another 120 degrees clockwise, you end up with the one to the right in Figure 4.14.

Now take any of the three spins (depicted in the left-hand vertical column in Table 4.3) and apply it to Figure in 4.13. Then follow that spin with another one of the three, from the top horizontal row. Thinking of Table 4.3 as akin to an addition or multiplication table, for example, you might have first selected a 120 degree spin, R_{120}, which would yield the

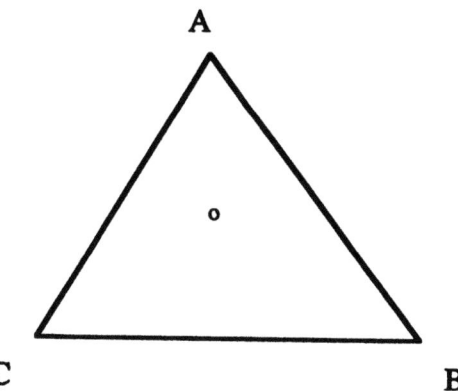

Figure 4.13 An equilateral triangle with a pinpoint at its center.

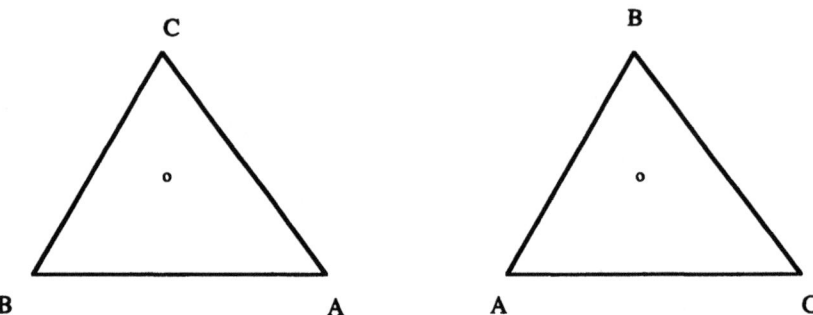

Figure 4.14 Spinning the triangle of Figure 4.13 −120 and 240 degrees clockwise

triangle whose position is on the left in Figure 4.14. Suppose you now spin that resulting triangle by 240 degrees, i.e., R_{240}. Then what figure do you end up with? You end up again with the triangle in the position indicated in Figure 4.13. That position is the equivalent of not spinning at all, or of spinning the figure 360 degrees about the center.

We can summarize what we have discovered by the following:

$$R_{120} * R_{240} = R_{360}.$$

This information is depicted in Table 4.3 by locating the first element, R_{120} in the vertical column on the left (the first deed performed) and by locating R_{240} in the horizontal row on the top. The box that describes the result, R_{360} is at their intersection (first column, second row of the table).

What is of interest from the point of view of same/different is that though everything in Table 4.3 is different from everything in Table 4.2 (the elements being classes of numbers in Table 4.2, while they are rotations in Table 4.3, the operation, ⊕ being a form of addition in Table 4.2, while * in Table 4.3 is something like "followed by"), the two structures are identical. They are *different* from the point of view of what they do

Table 4.3 Summary of the three spins of the equilateral triangle of Figure 13.

*	R_{120}	R_{240}	R_{360}
R_{120}	R_{240}	R_{360}	R_{120}
R_{240}	R_{360}	R_{120}	R_{240}
R_{360}	R_{120}	R_{240}	R_{360}

within each system but the *same* from the point of view of how they influence each of their respective systems.

Another Isomorphic Structure: Mack Truck Pulls Kiddy Car

In the above illustration, we have chosen an example of isomorphic structures that may appear a bit technical and belabored. Over the years, it has been acknowledged, however, in some form or other in mathematical curricula at a number of different levels of sophistication. By contrast, the example we now choose is so much simpler than the previous one that it is rarely seen for what it is worth—another striking example of isomorphic structures. In fact, the concept of isomorphism pervades mathematical thinking at the most elementary level. Consider, for example what most people do when they add numbers mentally that end with 0, such as:

$$\begin{array}{r} 430 \\ + 260 \\ \hline \end{array}$$

The usual fare is to ignore the zero, do the addition, and then append a zero at the end. In fact, it is possible to think of what we are doing schematically and in slow motion. We are in a sense focusing in two different directions. We can think of all the numbers ending in zero as being grouped together. We then associate each of them with an "alter-ego" in a second grouping for which the zero is lopped off. We then add the numbers in the second grouping. Once we have their sum, we associate that sum with a number in the original grouping (by appending a zero at the end).

One way of conceptualizing what we are doing is to realize that by "shipping" the numbers in the left oval to those in the right of Figure 4.15 we are dividing each of the original numbers by 10 (or multiply by 1/10). We add them together in the right hand oval, and shift them back to the left-hand side (essentially multiplying the sum by 10). The scheme (in four steps) is suggested in Figure 4.15.

Thus, 430 · (1/10) moves us into the right hand oval in Figure 4.15. 260 · (1/10) also moves us into the right hand sketch.

We get their sum in the right hand oval (step 3) and then move back into the left-hand one (step 4) by essentially multiplying by 10 as follows:

10 · { [430 · (1/10)] + [260 · (1/10)]} = 10 · (43 + 26} = 10 · 59 = 590.

It is essentially the application of the distributive property that accounts for why the short-cut works. The above diagram, however, provides a visual sketch of what is involved in the concept of isomorphism. That is, there is a shifting of the problem from the left-hand to the right-hand sketch, and then once the easier calculation is done in the new terrain, it is transformed back into the original one. The concept of isomorphism is in one sense simpler, but in another more subtle than in the one of the equilateral triangles and clock arithmetic. What we are essentially saying is that the system of addition for numbers ending in zero is different from that for addition for their counterparts that do not end in zero. If, however, we are smart about associating numbers from one system with that of another, these differences are not significant. In this way, we can capitalize on the fact that *dissimilar* systems are "essentially the same" from some point of view. (See Appendix B for a portrayal of the phenomenon of Figure 4.15 in terms of a fanciful story about "alter egos").

Before moving to yet one more way in which same/different functions in a mathematical realm, we might perhaps anticipate that the reader's reaction to what we have done in this section could justifiably be: We have used a Mack truck to pull a kiddy car! After all, to invoke the concept of isomorphism as an explaining mechanism for adding numbers that end in zero is to complicate something that is so simple it needs no explanation (or if one is offered, it should require much less baggage than we have brought to the situation).

We have, however, focused on an example that is overly simple precisely because we did not want the details of calculation to obstruct the

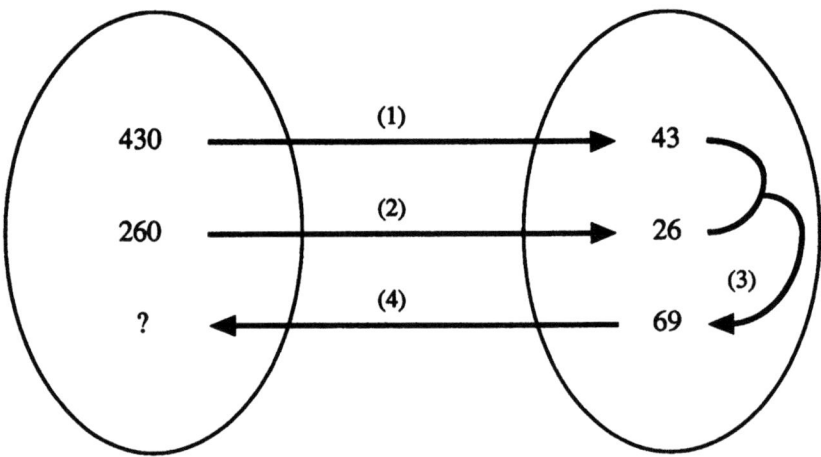

Figure 4.15 Depicting a scheme for simplifying addition of numbers ending in zero.

conceptual abstraction we were building. In order to appreciate the power of the distributive property, it might pay to return to the discussion of it at the beginning of chapter 1. Then the reader may wish to see how the concept of isomorphism is deeply implicated in the most general formulation of the property.

Unearthing Implicit Ingredients of Comparison
The above sections are intended to be a whirlwind review—by using examples in a relatively nontechnical way—of the conceptual centrality of similarity as a mathematical prize. In this section, we will be applying the concept not to one structure (with many different embodiments), but to an array of problems or structures in relation to each other. By doing so, we will make salient that problems are not only to be solved but can be compared with each other in a variety of ways, ranging from our personal attachment to them, to their cultural roots, to their mathematical connections. Bringing to the surface the concept of similarity that is usually handled subcutaneously is an important step in viewing mathematics from a more humanistic perspective. We will apply this concept again in the next chapter when we seek to broaden the relationship of mathematics to the real world. Equally important however, is the articulation of a language for similarity and difference. It will make it possible for us to speak about the phenomenon in a way that exposes its more fine-grained texture. We turn to that issue now.

As a prerequisite for the subsequent discussion, we begin with what may be a mundane observation. That is,

"Any two objects, thoughts, phenomena in the universe are similar."

If this statement seems peculiar, it is because when we claim that any two things are similar or more strongly the same, we are engaging in an ellipsis. Something important has been left out. That is:

"A is the same as (or similar to) B" is a short-hand way of expressing a more complex meaning. What we mean to say is:

"A is the same as B WITH REGARD TO C."

Frequently the C is so obvious that we take it for granted. Sometimes it is so obvious that we not only take it for granted, but we do not consciously notice it at all. From an educational perspective, it is sometimes worthwhile to be able to have the "Cs" become part of the conversation. To see the infinite variety of Cs enables us to increase our perspective on the world and invites us to see more of the world from the viewpoint of others.

What are some of the Cs that we might invoke—especially in relation to the mathematical experience? The following three problems will provide the impetus to analyze this question further.

Here are the problems (the second of which should look familiar, but it is repeated here for the convenience of comparison):

(1) The staff of P/L Press enjoys a winter skiing retreat every January. That is a time for bonding, but it is also a time for adventure. The ski lift moves at a steady pace, and Pablo, looking at markers along the way which indicate how far up the lift has traveled, notices that the lift moves one mile every four minutes.

Looking for "adventure of a new and different kind," Pablo asks the operator at the top if he would arrange for a special trip down at the end of the day satisfying the following condition: The constant speed on the way down should be such that the average speed of the entire trip (up and down) is twice the speed of the trip up.

Do you think the operator will be able to satisfy Pablo?

What should the rate of the downward journey be in order that his request be honored?

(2) Mordecai, a member of the advertising staff of the P/L Publishing Co. of NYC is enjoying his winter vacation at the Org ski resort in Orono, Maine. He is with a group of colleagues from the advertising staff of the company. His friend Pablo from the editorial staff of the same company was not able to join him when he left for Org, since the entire editorial staff was facing a deadline to edit a ground-breaking manuscript on the philosophy of mathematics education. Mordecai decides to make a long distance call to Pablo in NYC to tell him that Org is the most magnificent ski resort he's ever attended. Pablo decides to get the boss of P/L Publishing to offer a bonus to the entire staff for completing the editing job in two days. He gets it, and they are successful in completing the task in a couple of days. He and several of his friends from the editorial staff drive to Org to join the advertising staff. Along the way to Maine, Pablo and his friends see a Mercedes ambulance that passes them along the highway coming from the other direction. They pass at a point along the road that they notice a billboard advertising free ice cream. They comment on the Mercedes ambulance because they have never seen one like it before. Though they are tempted to stop for the ice cream, they are more interested in getting to the resort quickly than in stopping.

When Pablo arrives at Org, he finds out that Mordecai and the entire advertising staff have left the ski resort because their companion Chul Hong has broken his leg and needed to see an orthopedist whose office is located in NYC next door to the P/L Publishing Co. (The staff of P/L has an HMO insurance plan that would not reimburse its members for treatment other than at the NYC office unless the medical emergency was a life and death matter—which this is not).

The next day Pablo calls Mordecai who is back at work in NYC. He asks Mordecai how they got home with Chul Hong and his broken leg. Mordecai tells him that they were able to make a special deal with an ambulance that was about to head to NYC anyway, since the driver was planning to visit his sister there.

Furthermore, Mordecai tells Pablo that he thought he passed Pablo's car along the highway heading towards them when they were in the ambulance, but it was just a passing thought. In discussing the possibility, it turns out that Mordecai noticed the free ice cream bill board sign at the same time that he recalled passing the car that might have belonged to Pablo.

When they discuss the matter further, they both recall a fact that they find astounding. That is, it turns out that each group left their respective place at precisely the same time—heading towards each other's destination.

Furthermore, since it was such a surprise to see the free ice cream sign, both groups noticed how long it took to get to their respective destination after passing it. Mordecai said it took them nine hours to get home after seeing the sign, and Pablo told him that it took four hours to reach Org after seeing the sign. Furthermore, they know that the distance between NYC and Org is 300 miles.

At what rates were each of the two vehicles traveling? (Have patience. We will see this one again).

(3) About a month after they have arrived back at the P/L publishing Co. in NYC, Mordecai and Pablo find out that plans have been made to renovate some old offices for new staff who will be arriving in several months. The administration mentions that though they have funding for the room renovation and for a new computer for each incoming staff member, they do not have the resources for building and installing bookshelves. Hearing the sad news, Pablo and Mordecai volunteer their services to put bookshelves up in all the new offices. At first they do the work separately. After they have completed several jobs, they decide to collaborate. Pablo mentions to Mordecai that it took him four more hours to complete each job when he worked alone than when he collaborated. Mordecai looks back at his diary and he realizes that the situation was much worse for him. It took him nine more hours to do the job when working alone than when they collaborated.

The chair of the social foundations department overheard this conversation as he passed by the open door of the office that Mordecai and Pablo were renovating. He did not fashion himself being much of a math whiz, but he thought it would be fun to try to figure out how long it took them to complete the job when they worked together.

Do you think he was able to do it? Can you?

Now before either fainting or taking out a pad and pencil to try to solve or clarify anything, consider the following questions:

"How would you compare these three problems?" "Which ones seem most similar to each other and why?" "What are some of the similarities and differences?"

Of course a reasonable answer might be that they are all the same by virtue of the fact that they are all word or story problems. This realization may in fact generate another similarity among all of them. That is, they may all precipitate the same emotional reaction. They may all seem contrived; they might all be cause for stomach cramps; they might, however, sound like an enticing adventure; they might seem similar in that you would like to ask a slew of questions about each of the problems and of the key characters in them rather than meet the demand of the problem.

My experience has been that most people who feel a degree of comfort with problems of "this sort" (which itself is an interesting locution), claim that (1) and (2) are similar in that they fit a category that can comfortably be labeled as "distance problem." They would say that problem (3) is different from the other two in that it is a different sort of problem, dealing with work rather than distance.

By categorizing problems according to their "types" (as in distance vs. work), we have something of a *surface* similarity between problem (1) and (2). They have the earmarks of similarity by virtue of common language. The problems are the same in that they both speak of distance, speed, and time in contrast to problem (3), which uses time, but speaks of jobs being done.

Alternatively, there is an interesting similarity between problem (2) and (3). Though they are different in type, they both seem to be dealing with time in the same way. In both cases something takes "four hours more" and "nine hours more."

Despite the similarity of "type," between problems (1) and (2), they are different with regard to information conveyed. In problem (1) we do not know how far the ski lift travels, which may make it difficult to seek an answer to the question. Not only do we lack the necessary information from an explicit point of view, but we seem unable to figure it out from other information in the problem. In problem (2), however, we do have the relevant information of distance stated explicitly. It is 300 miles.

Analysis of this kind implies that, although we have not yet begun in any explicit effort to see the problems as something to solve, we have in fact begun to move in that direction. After all, in the absence of any demand (such as asking one to figure out an average round-trip speed that is twice the rate of the speed up the lift), it would not be of much relevance to point out that we do not know how far the lift travels on its one-way journey.

Let us move in the direction of solution for the purpose of seeing how that might transform the way in which we speak of similarity and difference among the problems. There are of course many ways to understand what each problem is about and also to create a scheme for solution. Not only algebra, but trial and error, graphing, asking for help from an "expert" to get us started are all legitimate ways of moving. I will not be overly concerned in the subsequent analysis with such details. Rather I wish to paint a picture of what might evolve in the way of thinking about similarity and difference if we follow some particular course of analysis. Other analyses might yield different categories of same/different, though I prize the ones we will end up with on the grounds that they open up some unusual territory that has wide applicability in other problem/solution contexts.

The first problem is actually one that caught Einstein's fancy. The various ways of analyzing the problem are not technically difficult, but we are put in the position of coming face to face with intuitions that must be reexamined. As a start, if Pablo travels one mile every four minutes, then he is traveling at a rate of 15 miles per hour. If the operator is to send the ski lift down at a rate such that the average round-trip rate is double the rate up, then how fast should Pablo be traveling on the way down?

A first intuition might be that since he wants to double the rate, he needs to come down at such a rate that the round-trip rate is 30 miles per hour. Well, if he goes up at 15 miles per hour and he wants the average round-trip rate to be 30 miles per hour, then it seems obvious that he needs to come down at 45 miles per hour. That is, to achieve an average of 30 miles per hour, it stands to reason that we compromise by choosing 15 miles per hour up and 45 miles per hour down. The average of 15 miles per hour and 45 miles per hour is 30 miles per hour.[15]

Now look at the situation a bit differently. If Pablo wants to come down at a speed such that the round-trip rate will be double his speed up the lift, then one way of looking at the problem is to realize that he must travel twice the original distance in the same time it took him to go up. If that is the case, then how much time does he have to come down? It seems that the answer is zero time down. Is that possible? Maybe this problem is one that has no solution despite the fact that it appeared to have a straightforward one.

There is an interesting insight that this "commonsensical" approach reveals that is hidden by a comparable algebraic one. That is, we can reach the conclusion that the actual distance of the ski lift is not really relevant. Hence a phenomenon that seemed important, and in fact distinguished one problem from another at first, turns out upon a modicum of reflection not to be a distinguishing feature after all.

Now look at problem (2). How do we think about it? Again, there are many ways, but in order to get at a bigger picture, we are going to move along algebraically and in a way that does not express alternative modes of operating nor reflect the evolution of one's thinking. We do not know how long it takes for the cars to meet, so let us label that period of time t. Then we do know that Pablo's group traveled for four more hours: (t + 4). Mordecai's group traveled for nine more hours: (t + 9). They traveled at different speeds, however. Therefore let us denote the speed of Pablo's group as s_1, the speed of Mordecai's as s_2.

Since the distance is 300 miles in each case, and the product of rate and time equals distance, we have

(i) $s_1 \cdot (t + 4) = 300.$
(ii) $s_2 \cdot (t + 9) = 300.$

Figure 4.16 is a sketch that illustrates equations (i) and (ii)

Each of the equations (i) and (ii), focuses on the travels of each of the two groups (the top of Figure 4.16 is Pablo's, the bottom is Mordecai's) one at a time. A is the point at which they meet in each case.

We have almost enough information to do the necessary algebraic mystification. However, since there are only two equations but three unknowns, we need to inject one additional bit of information. The sketch of Figure 4.17, suggests that one way of doing so is to combine the travels of each of the two groups in one equation. That is, Pablo's group travels part of the way to the meeting point A, and Mordecai's group the other part of the way from the same meeting point.

The sum of the two separate distances to the meeting point is thus 300 again. So:

(iii) $s_1 \cdot t + s_2 \cdot t = 300$

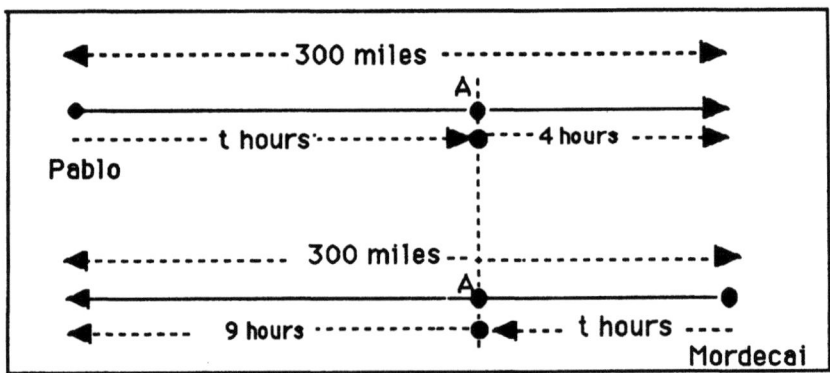

Figure 4.16 Depicting the travel of Mordecai and Pablo in problem 2.

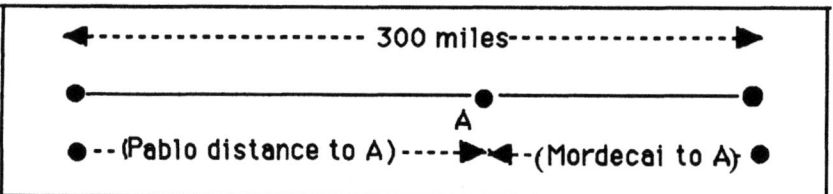

Figure 4.17 Depicting additional information for problem 2.

Solving equation (i) and (ii) for s_1 and s_2 respectively, we get:

$$(i') \quad s_1 = \frac{300}{(t+4)}$$

$$(ii') \quad s_2 = \frac{300}{(t+9)}$$

Placing these values of s_1 and s_2 into equation (iii), we get:

$$(iii') \quad \frac{300}{(t+4)} \cdot t + \frac{300}{(t+9)} \cdot t = 300$$

Dividing by 300 (or multiplying both sides by 1/300; or canceling the 300s, depending upon whether you are over 50 years old, went to public school or know the songs of Tom Lehrer), we get:

$$(iv) \quad \frac{t}{(t+4)} + \frac{t}{(t+9)} = 1$$

If we solve equation (iv), we end up with +6 or −6 as a solution. Given the nature of the problem, we select +6 as the solution for t, which in turn tells us that the velocity of Pablo's group was 30 miles per hour and Mordecai's was 20. (Someone did a lot of smelling the roses on the trip.) We have perhaps gone overboard in some algebraic analysis, but there are many important points that flow from this analysis as we re-adjust our lenses and retrieve the forest from the trees.

First of all, look at equation (iv) once more in comparison with equations (i), (ii), and (iii) that generated it. What relevant information appears to be missing? The "cancellation" of the 300 at the end of this ordeal is revealing. In some important sense the distance that separated the cars at the beginning seems to have been washed out at the end. Apparently the final solution would then be the same if the distance traveled were 1,000 miles or one mile! Why does that happen?

Before thinking about this question a bit more, if we return to an early observation that was based more upon the statement of the three problems, we uncover a fascinating reversal of similarity and difference. At first we observed that problems (1) and (2) were the same with regard to "type" (both distance). Further reflection led us to believe, however, that despite this commonalty, they were essentially different. That is, problem (1) appeared to lack pertinent information which problem (2) exhibited—the distance of the ski lift and the distance between the two parties, respectively. This was our inclination as we entertained in an intuitive sense what a solution might entail.

The above analysis reveals an interesting reversal. Although there is a difference with regard to information about distance conveyed in both problems, it is not that the first problem lacked the appropriate information, while the second had it. Rather, what the first problem lacked and what the second problem included were both irrelevant in terms of finding a solution. This last observation, however, has deep consequences. In what sense is the actual distance irrelevant in problem (2)? Though the distance traveled does make a difference in terms of analyzing the actual rates that each party travels, the amount of time needed for them to meet is the same regardless of the distance. Why is that?

The answer to the question is suggested by the actual form of equation (iv). Look at that equation again. It is the one that captures the essence of problem (2). What is that essence? A clue to the answer is further revealed by looking once more at problem (3): a problem dealing with work and not distance at all. Without engaging in a detailed analysis and also realizing that there are many other ways of analyzing problem (3), it turns out that equation (iv) is one way of placing that problem in an algebraic format. That is, equation (iv) is associated not only with problem (2), but with problem (3) as well! It is precisely the sort of equation we might have set up without going through any intermediary equations if we were faced with problem (3).

What does this mean? While the existence of an identical equation to capture the essence of two different problems is not automatically an indication that the problems themselves are essentially the same, for it could be a coincidence, it is suggestive. It leads us then to wonder about the actual relationship between problems (2) and (3). Could they be "essentially the same"? Is there a way in which we might think of problem (2) as a problem that could be construed in terms of work and not distance? Asking the question is a hair's breath away from supplying the answer in this case. Suppose we think of the distance between the two points (call them P and Q) as a "job to be done." Then one car (the one from P to Q) can complete the whole job alone in four more hours than it takes for them to do it jointly, while the other car (from P to Q) can complete the whole job alone in nine more hours than it takes for them to do it together. The general point is that problem (2) is in fact essentially the same as problem (3), though the language used is different in the two problems. What this analysis suggests is that "work" problems may be essentially the same as "distance" problems, though the similarity is carefully camouflaged.

"With Regard To": Educational Excavations
We chose problems (1) through (3) in the previous subsection carefully in order to bring out some rather undervalued ways of speaking about similarity in mathematical thought. As with much of the rest of this book, however, our intention is suggestive rather than didactic. The general purpose here is to indicate how differently we think when we move from problems (and situations) viewed individually to seeing them in a comparative mode. The ones selected here provide a powerful language to discuss a host of issues related to the concepts of same and different.

We began by exploring possible emotional reactions to the collection of problems and moved from there to ways of relating problems that bypassed a focus on solution. Eventually, however, we reexamined the concept of similarity and difference for these problems in light of what we found out as we moved in progressive stages towards solutions.

As we moved along the spectrum of observing problems in a way that severed them from solution to examining them in the context of solution, we came upon some interesting interpretations of Cs. That is, we asked: What might be some relevant Cs in the following assertion?

"A is like B with regard to C."

What we noticed was that one possible C is "surface similarity." It appeared at first that problems (1) and (2) were the same from the point of view of a surface category. So, we have:

C_1: "surface similarity," expressed by similar language used in the statement of the problems.

Further exploration of problems (1) and (2), enabled us to see that one was solvable and the other was not. Thus, despite the surface similarity, they were quite different from each other from an important conceptual point of view.

Although in the early stages of thinking about them, problems (2) and (3) seemed to differ significantly, they in fact were *conceptually* similar. This appears to be a category that is not necessarily revealed in the language used to describe the problem, and in the most extreme case is exemplified by the concept of isomorphic structures, described in two previous sections of this chapter. This issue introduces the concept of "deep similarity."

> C_2: "deep similarity" indicates that problems, which may appear to be different from each other, are in fact conceptually similar.

It was in the act of focusing upon the algebraic solutions to problems (2) and (3), that we suspected that their similarity might be deep. The solution to problem (2), which was seen as a "distance" problem had a *form* that was identical to that of the solution to problem (3), a "work" problem. The equation we finally arrived at in order to accommodate the detailed information of problem (2) was the same as that of problem (3). Thus, we have yet another way of characterizing C's. That is:

> C_3: "formal similarity" between and among problems by virtue of the nature of their solutions.

At some stage in comparing problems, we may notice that they are similar not with regard to their surface, nor with regard to something deep, but rather with regard to the *form* of their solutions. There is no guarantee that similarity of form indicates anything other than coincidence. In this case, however, we used the similarity of form to investigate further in an effort to reveal a deep similarity. It is by "comparative education" that we can begin to see some unusual similarities and differences. The vocabulary of the C_i s enables us to carry on this sort of conversation. It is one that not only invites our students to reveal their evolving perceptions and competing attitudes about problems and problem solving, but also provides a way of talking about the mathematical concepts in a richly textured way.[16]

One interesting consequence of the language of the Cs is that it shares a way of thinking comparable to linguistics. Noam Chomsky popularized this way of talking with regard to the acquisition of language. As a matter of fact, we might venture that, after carrying out the analysis that relates problem (2) to problem (3), we are on the verge of discovering that distance and work problems in general, not just the ones we selected, might in fact share a deep similarity.

Furthermore, it is a kind of language that encourages us to revisit some of the more standard and informal ways of talking about similarity that we used to characterize mathematics at the beginning of this section. We might also want to revisit some of the substantive matters we discussed with regard to human nature. The connections are not automatically generated based upon mathematical analyses of the sort we have discussed with regard to same/different. Analyses of the sort we have been advocating, however, provide the opportunity for raising the conversation from that of mathematical problem/solution to metamathematical/humanistic categories. Once again, Figure 4.6 invites us to transform what are originally mathematical problems and solutions into problems and situations that emphasize human agency.

We should stress that what we have explored is a case study of sorts. In playing around with the relationship among three problems, we have come up with a kind of language to use that was both generative and descriptive of our quite specific inquiry in comparing them. It appears, however, to have promise for the comparative investigation of mathematical problems, situations, and solutions in a wider realm. This is just the beginning of such inquiry, however. There are obviously many (perhaps infinite) ways of depicting the Cs in the assertion "A is like B with regard to C." Among the Cs we might find worth investigating are ones that describe such categories as aesthetics, emotionality, cleverness, potential, obtuseness, surprise, and a host of other qualities.[17]

Finally, what we have discussed here with regard to the schematic "A is the same as B with regard to C," has much in common with what we will explore in chapter 5 under the heading "Language and Metaphor." There our concern, however, will be more on how we might expand the concept of "application" of mathematics to the real world. In addition, we will be less focused on problem and problem solving and more concerned with seeking forms of connection between what is generally considered to be mathematical verses nonmathematical domains.

Part III
RECONSTRUCTING REAL-WORLD CONNECTIONS

Chapter 5

An Enhanced View of Connecting Mathematics With the Real World

There are more things in heaven and earth, Horatio, than are dreamt of in your philosophy.
—Shakespeare, *Hamlet*

We now turn to the second main theme of this book: connecting mathematics with the real world. That focus is intertwined with problem solving and is generally associated with a desire to make use of applications. Some minor excavation of the concept of applications reveals that though it may be implemented in a subtle manner, for the most part what is meant by "application of mathematics to the real world" means essentially that mathematical models are created that can be used to explain and predict solutions to real-world problems. We explore that concept in this chapter and will offer both expansions and alternatives to it. In the final chapter, we take another look at the concept of "real" itself, and in so doing, will consolidate our criticisms of the themes of problem solving and connecting mathematics with the real world.

The Established Concept of Connection

Frequently, the pedagogical strategy for arriving at a model is a gentle and inductive one, and it is sometimes not made explicit. Thus, students may be taught to create tables of information in order to arrive at some functional relationship that is eventually expressed algebraically. Alternatively, some geometric model might be sought in order to isolate the relevant variables in the problem.

The justification for using real-world problems is frequently made on grounds that they provide motivation for mathematical inquiry that might otherwise be considered as isolated from experiences in the world. In addition, many educators believe that real-world problems have a "concrete" quality that provides a transition towards concepts that will be handled in some more abstract (mathematical) way down the road.

The notion of modeling as a concrete connection is depicted in *The Standards* as follows: "As in the earlier grades, teachers in grades 9–12 should introduce a new topic by exploring appropriate concrete representations in which students recognize that the exploratory activities embody the mathematical topic. This helps establish modeling connections, which can be further strengthened by an instructional approach that encourages multiple methods of solution for any given problem" (National Council of Teachers of Mathematics, 1989, p. 147).

Borrowed from a sketch in *The Standards*, the concept of model is elegantly schematized by Dossey (1996) in Figure 5.1.

In its most refined sense, what we seek is some sort of correspondence between elements of the real world and associated mathematical ones. Activity in the real world is thus associated with mathematical operations. The scheme suggests that we both begin and end with some "real-world" phenomenon and that we invoke a variety of mathematical moves along the way. One clear implication of the use of a mathematical model is that the world is always a bit "messier" than the mathematics that models it. Thus, we need to figure out what is relevant and what is

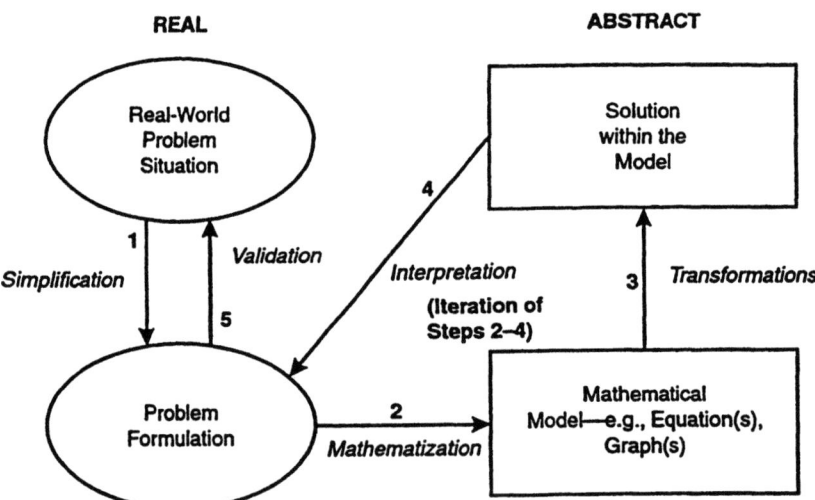

Figure 5.1 Standard scheme for a mathematical model of a real world problem. (From Dossey, 1996, p. 55, reprinted by permission of John Dossey and Heinemann, a division of Reed Elsevier Inc., Portsmouth, NH).

"noise" as we seek to find some correspondence between mathematics and the real world (depicted by "simplification" as an early step in Figure 5.1).

One of the first things students learn is how to determine and thus eliminate irrelevant information. For example, if we are told that there are two trains traveling towards each other at certain rates and that the engineer in the first train leaves at 1 p.m., and the second leaves at 5 p.m., we need not attend to the facts that the trains per se are the modes of transportation, nor that the time of departure is at a particular time of day (though the relationships between the times of departure will probably be relevant) nor that the driver is an engineer, nor what the gender of the driver may be, nor what the weather conditions are like.

We need not even pay attention to what may have motivated the people to take a train and to engage in this sort of adventure in the first place. Knowing what to eliminate is part of what is involved in making sure that one creates some appropriate mathematical model of a real-world problem.

To say that we need not take "irrelevant" information into account does not imply that the advocates of integrating mathematics lack an appreciation for subtle messages that are conveyed by the problems selected.

Recall the problem of Anne, the basketball player, that we discussed in chapter 1. It was most likely motivated in part by an attempt to confront, in a somewhat subliminal manner, a sexist attitude towards school athletics. "Suppose Anne tells you that under her old method of shooting free throws in basketball, her average was 60%. Using a new method of shooting, she scored 9 out of her first 10 throws. Should she conclude that the new method really is better than the old method?" (National Council of Teachers of Mathematics, 1989, p. 172).

This is a significant problem for students to think about since it invites them to see events in the real world from a probabilistic point of view. We cannot answer with any degree of certainty whether Anne's new method is superior to her old one, for it certainly is possible that Anne's new method resulted in a superior performance just on the basis of chance. Using probabilistic thinking, however, we can make a statement about the degree of confidence we might have in a conclusion.

It would seem difficult to criticize what we have portrayed in the above discussion as a basic way of linking mathematics with the real world. We might, for aesthetic or political reasons, seek and replace some of these connections with other fields that have not been represented as connecting sources. For some of the examples, practical problems may arise, e.g., their inappropriateness for a particular grade level. We might be concerned with the contrived nature of some of these examples, as is the criticism by many educators of the standard word problems that have been the bane of existence of many youngsters for a long time. Nevertheless, it looks as if the concept itself of connecting mathematics with the real world via appropriate problem situations is difficult to criticize.

We turn now to such a criticism. As in our earlier discussions of the relationships between problems and their solutions, we are not claiming that the alternative views portrayed below are more effective in some empirical sense of "what works better" for students. However, the vision of integration is—as in the case of problems and their solutions—a truncated one. What is needed is to appreciate how that truncation has taken place in order to generate reasonable alternatives that might be explored.

Alternative Conceptions of Connections

Imagine the following problem as an invitation to consider alternative conceptions of connecting mathematics with the real world.

> A close relative of yours has been hit by an automobile. He has been unconscious for one month. The doctors have told you that unless he is operated upon, he will live but will most likely be comatose for the rest of his life. They can perform an operation which, if successful, would restore his consciousness. They have performed ten such operations in the past and have been successful in only two cases. In the other eight, the patient died within a week. What counsel would you give the doctors?

In what sense does this problem represent something that is different in kind from the earlier one about Anne? They both deal with probability. They both involve some need to come up with a conclusion. The present problem, however, goes beyond conclusion to require a decision. Surely, however, more than decision making distinguishes the two problems. The current problem deals not only with problem solving of a mathematical nature but with ethics as well. We have to consider the possibility of supporting an event with a low probability. In addition, we have to ask not only what the chances are that the person will survive, but we need to confront the more painful question of what kind of life is worth living.

It is not only that the issue we confront is an ethical one, though this quality makes it a more interesting problem for many students who may see the use of mathematics in the real world as somewhat contrived. More generally, in order to make sense out of a real-world problem, mathematical consideration is one important dimension, but it is part of a larger cloth.

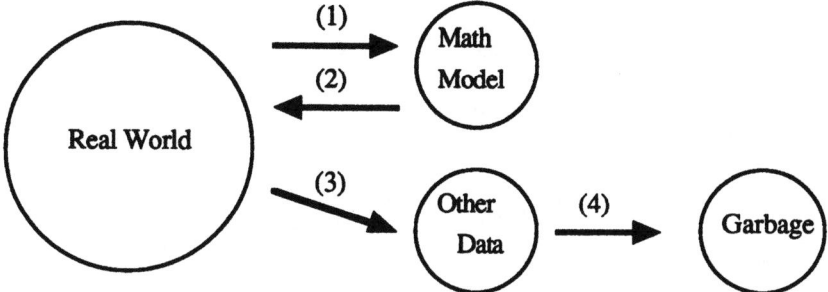

Figure 5.2 Making salient the discarding of "irrelevant" information in Figure 5.1.

We cannot come to an intelligent conclusion about what to do if we whittle away at the "irrelevancies" as we did in the example with Anne, the basketball player. The critical element here is that we are dealing with a person's life. We probably would want more rather than less information about the comatose friend. We might want to know what kind of life the person led. We might want to know what sort of living will, if any, the person left. We might want to seek additional medical opinions rather than accept the conclusion about what might or might not be the result of performing the operation. In short, in order to make a wise decision, having a lot more information would be more desirable than streamlining the real-world problem for the purpose of creating a workable mathematical model.

Let us now oversimplify Figure 5.1 with the intention later on of elaborating upon it, given the ethical dimension of the suggested variation on the basketball problem. A slight caricature of Figure 5.1 is depicted in Figure 5.2. It suggests that although we are aware of the excess baggage

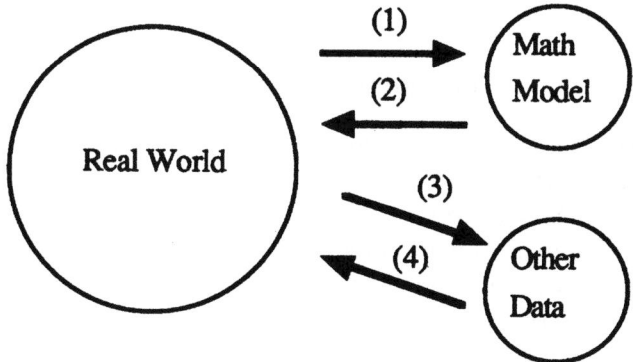

Figure 5.3 Seeking relevance of given non-mathematical information in relating mathematics to the real world.

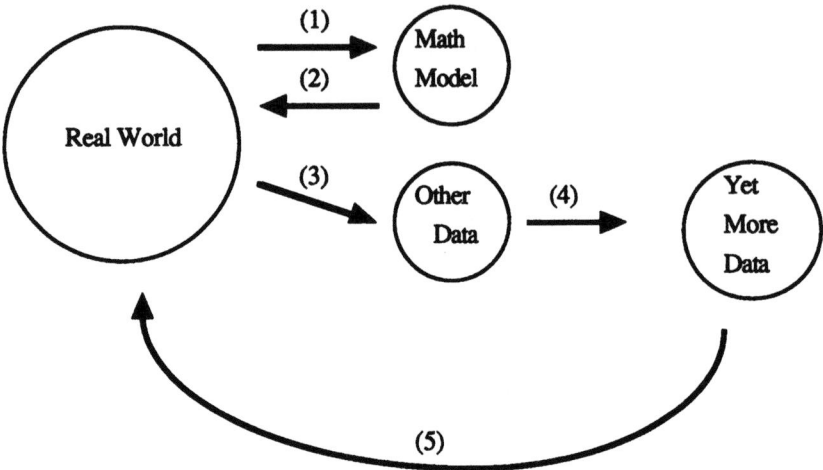

Figure 5.4 Searching for additional information of a non-mathematical nature in relating mathematics to the real world.

in Anne's real-world problem, we choose to ignore it completely. The revised problem with the comatose friend, however, adds a new dimension to the concept of connecting mathematics with the real world. We do have information that is not strictly speaking mathematical in nature, but we need to make use of this information as depicted in Figure 5.3

As we have indicated above, the situation is even a bit more complicated since we need to make use of "other data" that may already be available, but we also need to seek out new information in order to make a wise real-world decision—information involving values that we and our comatose friend hold about the nature of life. The more elaborate picture is depicted in Figure 5.4.

Yet Another Scheme

The scheme of Figure 5.4 begins to turn our glance in a different direction from the standard view of relating mathematics to the real world through the use of models. Still, there are remnants of the earlier scheme depicted in Figure 5.1. The most significant remnant is that of mathematics as an element in the service of some real-world problem or phenomenon. As the ethical problem suggests, it is not the only remnant, but it is used for the purpose of advancing some pressing concern. What are other conceptions of relating mathematics to the real world that could suggest other purposes than those depicted in the above schemes?

A substantially different point of view is revealed as soon as we begin to relinquish a hold on mathematics that is rooted in a desire to see the field as totally different from other experiences in the world—as the deductive science par excellence, the science of necessary conclusions, or the field in which knowledge that is certain may be purchased to rejuvenate some other real-world phenomenon. With this acquiescence, we invite a point of view that asks a different sort of question. What is there in the way of thinking, experiencing, feeling that mathematics shares with other ways of experiencing the world? A scheme that acknowledges this perspective is depicted in Figure 5.5. Here we have an array of human qualities, concerns, interests that transcend specific disciplines and that are as much a part of the mathematical experience as well.

Here we have an array of human qualities, concerns, and interests that transcend specific disciplines and that are as much a part of the mathematical experience as well. Our purpose in this section is suggestive. Rather than presenting a deep analysis and rationale for the inclusion of each of these qualities, we will be more inclined to offer a point of view

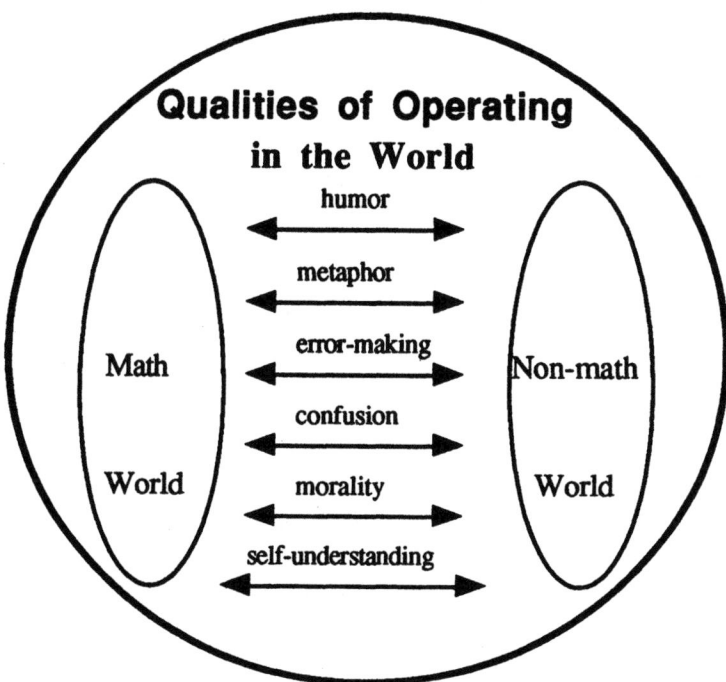

Figure 5.5 Humanistic connections that relate mathematics to the non-mathematical world.

that is anecdotal, personal, and compelling. It is, however, reminiscent of what we discussed in chapter 4 with regard to the basic human lens of "same/different." Here we are not so much looking at problems and subfields within mathematics in an effort to uncover essential similarities and differences from the perspective of problem and problem solving. Rather, we are choosing a lens that attempts to see how pieces of the mathematical experience derive from and compare with other fields and experiences—not from the viewpoint of models, but something we shall clarify after offering some examples.

Language and Metaphor: From "Beware" to "Be Aware"

Mathematics can be connected with language in numerous ways. There is, of course, the sense in which mathematics is itself a language. Having said that, many issues flow from this observation. In what sense is it that mathematics is a language? How does it compare with ordinary language? In what ways does it depend upon ordinary language. We forego the analysis of those questions directly. Instead we focus our attention more sharply on one piece of a connecting link between mathematics and language, a piece that is accessible in the absence of a strong philosophical backdrop. Reuben Hersh (1997), author of popular books on the nature of mathematics, reflects as a teacher and internationally known research mathematician upon the difficulty his students have had in learning mathematics during his long career.[1]

"Beware the double entendre" is a good slogan that explains, according to Hersh, a major reason that students have difficulty with learning mathematics. Ordinary language is not only filled with ambiguous meanings, but, he argues that even when there is no ambiguity in ordinary language, there is generally either no connection or a tenuous one between that meaning and the mathematical term.[2]

As an example of a tenuous connection, Hersh (1997) comments, "If I say 'I own a number of calculus books . . . ,' I don't mean *zero books*. . . . I don't even mean *one* book. . . . I mean two or more" (p. 48). He claims that he now understands that it was not mere ignorance that accounted for the comment many years ago by one of his students who asserted that zero was not a number.

Hersh offers a litany of other ordinary language expressions that are at odds with mathematical meaning: *adding,* which in ordinary language always leads to an increase in number, *difference,* signaling a comparison in ordinary language, but not necessarily subtraction, and *multiplication,* repeatedly adding so that one arrives at something that is bigger than what was initially the case.

Connecting Mathematics With the Real World

The connection between mathematics and ordinary language can be even more tenuous however in advanced mathematics, as Hersh points out: "In advanced mathematics, there's more linguistic confusion. Surds (absurd), irrational and imaginary numbers, singular perturbations, degenerate kernels, strange attractors—all sound dangerous, undesirable, things to avoid "(p. 51).

What, however, do we conclude from Hersh's observation of the ambiguity of meaning? It is true that the mismatch between mathematical and everyday meanings is significant enough to warrant our attention, and a disinclination to appreciate this observation may very well account for students' problems in appreciating mathematical meaning.

Nevertheless, there are concomitant issues that are either ignored or distorted by Hersh's program to clear up the intended *entendre*—in an effort to minimize ambiguity. These issues have deep consequences not only for students attempting to learn new bodies of knowledge, but for anyone trying to appreciate the nature of mathematical thought as well as its intellectual history. Thus, I would like to suggest the following complementary slogan to Hersh's "Beware the double entendre." It is "Be *aware of* the double entendre."

Precision of meaning is one thing. An appreciation for the evolution of ideas and the associated labor pains is another. The slogan "Be *aware of* (rather than *beware*) the Double Entendre" is intended to have an ameliorative rather than a dismissive quality with regard to the concept of double entendre. While Hersh has found out that some students have trouble understanding a concept like that of irrational or imaginary numbers because they seek association with words that "sound dangerous, undesirable, things to avoid," I have discovered that many are frustrated by a disinclination to take seriously the ordinary language equivalent.

Take the case of "negative number" for example. While "negative" surely fits the bill of sounding dangerous (unless of course it is associated with a biopsy), the Latin translation of that concept (which predated the English translation) was just as foreboding and perhaps more revealing. As we mentioned in chapter 2, these numbers were originally called *numeri ficti*— fictitious numbers. The implication here is not only that these numbers are dangerous, but that they really do not exist—or if they do, their existence is shrouded in mystery.

What can students learn not by dissociating from an English translation, but by embracing such translations with an historical and multicultural perspective? Perhaps the deepest lesson to internalize is that they are not fools if they do not immediately understand what the concept is all about. Not singly, but taken as a whole, words like "negative," "imaginary," "irrational," "complex" with regard to numbers signal something very important. They suggest that these concepts evolved against considerable resistance. Students may come to appreciate that in a quite deep sense, "ontogeny recapitulates phylogeny." If our students have trouble understanding how numbers are extended, then it may be solacing for them to know that they are merely experiencing the historical labor pains of these ideas.

Why should these ideas have had such a labor intensive birth? Why have they not just been accepted as reasonable extensions of existing knowledge? What does it mean to say, as Hersh points out, that mathematicians appreciate that zero may have meaning in the above context while ordinary language suggests the opposite? Who are these mathematicians who appreciate the meaning? Are we referring to those who gave birth to the ideas and found themselves walking on a tightrope, or are we referring to a twentieth-century embodiment of "mathematician?" Are there present-day mathematicians who would have difficulty with the concept of zero defining a number of real-world objects? Should there be?

Questions like the ones above are inspired by attending carefully to metaphors used in mathematical inquiry. In addition to referring to historically based metaphors, it is worth becoming aware of and appreciating the metaphors we all use in our efforts to understand, to solve, and to generate mathematical concepts and problems.

Metaphor is a concept we normally associate with poetry or other forms of literature in the "real world." In fact metaphor is so deeply implicated in all our thinking that we engage in variations of metaphorical thinking, even when we are not aware that we are doing so.[3]

Metaphor: An Example

A personal anecdote below illustrates how metaphor has functioned in my own thinking in the realm of easily accessible ideas. What started out as a doodle of sorts ended up with an awareness of the power of metaphor in mathematical thinking.[4]

Table 5.1 Doodling with some simple multiplication facts.

$$1 \cdot 3 = 3$$
$$2 \cdot 4 = 8$$
$$3 \cdot 5 = 15$$
$$4 \cdot 6 = 24$$
$$5 \cdot 7 = 35$$

Connecting Mathematics With the Real World 149

I was staring at the "data" of Table 5.1—data that I was barely aware I had created.

At first a number of simple observations barely entered my consciousness. I noticed, rather unenthusiastically, for example, something about how the numbers on the left-hand side of the equation in Table 5.1 increased pairwise. Thus 1 and 3; 2 and 4; 3 and 5 all differed by 2. Looked at from the point of their columns rather than rows, each element of the pair increased by 1 as the numbers progressed vertically.

Soon my awareness became more intense as I focused on the numbers on the right-hand side of the equation: 3, 8, 15, 24, 35. This was a kind of progression that I had met before. That is, the differences between the successive right hand vertical terms were not a constant. Rather they formed the progression: 5, 7, 9, 11 (e.g., $8 - 3 = 5$; $15 - 8 = 7$; $24 - 15 = 9$).

Though I had been trained to make this type of observation, since such differences (arithmetic differences) signaled something about the kind of equation (quadratic) that might generate the original pairs, I noticed an even more remarkable quality about the numbers on the right-hand side (3, 8, 15, 24, 35). I saw them not as what they actually *were* but rather what they seemed to be *trying to become*. With just a small amount of squinting, these numbers were all almost something else—and furthermore they all missed being that something else by the same amount—one unit.

In fact, each of the numbers on the right-hand side is almost a perfect square. I became intrigued with the metaphor of numbers "striving" to become a square. This idea is depicted in Table 5.2

I moved in two directions almost simultaneously. One impulse was to do a What-If-Not on the original list. If something so remarkable could occur with pairs that differed by 2, what might happen if they differed by some other fixed amount? Would there still be a "striving" metaphor to pursue? A second impulse was to try to understand why this was happening. In this case, my inclination was to seek some sort of geometric rather

Table 5.2 "Striving to be squares" as a way of seeing Table 5.1.

$1 \cdot 3$ *almost* equals 4
$2 \cdot 4$ *almost* equals 9
$3 \cdot 5$ almost equals 16
$4 \cdot 6$ almost equals 25
$5 \cdot 7$ almost equals 36

Table 5.3 Modifying the doodling of Table 5.1.

$$1 \cdot 5 = 5$$
$$2 \cdot 6 = 12$$
$$3 \cdot 7 = 21$$
$$4 \cdot 8 = 32$$

Table 5.4 "Striving to be squares" as a way of seeing Table 5.3.

$$1 \cdot 5 = 5 \longrightarrow 1$$
$$2 \cdot 6 = 12 \longrightarrow 4$$
$$3 \cdot 7 = 21 \longrightarrow 9$$
$$4 \cdot 8 = 32 \longrightarrow 16$$

than algebraic understanding—the reason being that multiplying pairs seemed to be speaking to me about areas of rectangular shaped figures. My desire to seek something geometric was motivated in part by the realization that, while algebra might provide some formal justification, it would not enable me to "see" what might have motivated the algebraic moves. In line with my first impulse, I sketched Table 5.3.

What happened to the "almost squares" on the right-hand side of the equation now? There were many ways to see what might be happening. Instead of 5, 12, 21, 32, I imagined the "strivings" as 1, 4, 9, 16. So perceived, instead of striving to become a large number that was a square, they each strove to become a smaller number that was a square. Table 5.4 describes the situation.

There appears to be a pattern emerging between each number preceding an arrow and the square numbers to the right. The pattern in the $1 \cdot 3$ array of "missing by 1" seemed to be destroyed, however, as seen in Table 5.5.

Table 5.5 Depicting the "short-fall" of the "striving" of Table 5.4.

$$1 \cdot 5 = 5 \xrightarrow{4 \text{ off}} 1$$
$$2 \cdot 6 = 12 \xrightarrow{8 \text{ off}} 4$$
$$3 \cdot 7 = 21 \xrightarrow{12 \text{ off}} 9$$
$$4 \cdot 8 = 32 \xrightarrow{16 \text{ off}} 16$$

Table 5.6 Revisiting the "striving" of Table 5.4 in search of a constant "short-fall."

$$1 \cdot 5 = 5 \xrightarrow{4 \text{ off}} 9$$
$$2 \cdot 6 = 12 \xrightarrow{4 \text{ off}} 16$$
$$3 \cdot 7 = 21 \xrightarrow{4 \text{ off}} 25$$
$$4 \cdot 8 = 32 \xrightarrow{4 \text{ off}} 36$$

Was there some way of retrieving the pattern of "striving" that was depicted by a loss of 1 in each case? A small amount of fumbling around revealed that if I shifted all the squares up so as to think of 9 and 16 rather than 1 and 4 as the first two squares, we would arrive at a very similar notion of "striving" once more. The pattern in Table 5.6 emerged to fit the bill.

Before moving to my second impulse, a search for a geometric depiction, I noticed that the "striving to be squares" missed by 1 in the case of the original doodling in Table 5.1 (1 · 3 and following) and by 4 in the case of the new doodle of Table 5.6 (1 · 5 and following). I wondered about the significance of the numbers 1 and 4, and I also then became aware of the fact that my concept of "striving" was really two concepts rolled up in one: "striving for something other than what it is," and "missing that goal by the same amount for each pattern."

How might I use these observations to gain some geometric insight. I chose the third entry of my first array, the 3 by 5 product from Table 5.2, and sought a geometric representation. How could I then represent a 3 by 5 array by using the same metaphor as before—"striving" to become a square, or being an "almost square"?

One rather simple scheme is suggested in Figure 5.6. Skipping the details, it seemed clear to me that such a transformation of rectangles to

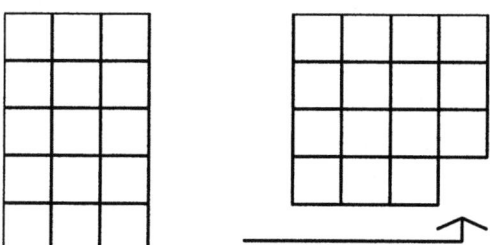

Figure 5.6 Geometrical depiction of "3 · 5 almost equals 16" from Table 5.2.

"almost squares" (missing by 1) works in all the cases of the first array. There was nothing special about 3 by 5. An exploration of 4 by 6 or 8 by 10 reveals the same "almost square" that is arrived at by moving a bottom line of the rectangle.

Where to from here? I will spare the reader a painstaking analysis and get right to the quick. As a result of this doodling, and inspired by an awareness of metaphor as well as an application of What-If-Not thinking, I had been able to reduce the problem of multiplying any two integers to that of manipulating only squares.[5] In a sense, though calculators have paved the way (perhaps) for not requiring memorization of multiplication tables, I was able to show that such tables were highly redundant. That is, it would be almost enough to have tables of squares of numbers rather than standard multiplication tables if we want to find the product of any two numbers. We can always represent the product as the difference of two squares.

The issue of "success" here is less significant than our coming to an awareness of the metaphors that drive our thinking. Even when they do not work out immediately, invoking even a mild What-If-Not perspective on an emerging metaphor has the potential to reveal depth that did not appear on the surface.

It is a significant educational agenda to become aware of and share the metaphors and images we use when we teach and learn and to wonder about how metaphorical thinking functions in our thinking in general. In chapter 1, we saw the hypocrisy and fear that results when we hide the metaphors we use (as in the introduction to the course in finite dimensional vector spaces). We also discovered how some of this fear may be expressed and handled in a potentially therapeutic way when images and metaphors are not only expressed, but created as part of the act of learning (as in the student's description of the cold hand of mathematics in that chapter as well).[6]

Before closing this sub-section, we turn to an interesting twist about the concept of metaphor and its functioning in mathematical discourse.

Metaphor: Another Ironic Twist

There is a sense in which mathematical thinking connects mathematical ideas in a metaphorical way with those that everyone experiences in the context of daily living. It is not the pale coin of *applying mathematics to the real world*. Such a view assumes that the real world and mathematical thinking are in fact separate entities and that there is an occasional opportunity to seek their linkage. What I have in mind is that mathematical language offers metaphors for understanding the most fundamental qualities of human existence. Such a point of view stands on its head the connection between ordinary language and mathematical thought, that we transport metaphors from ordinary language to mathematical experiences as if the former had a prior and independent existence from the latter. The potential for reversal of this direction was unveiled in the early twentieth century by Keyser (1916), chair of the mathematics department at Columbia University at the time. He depicts this connection as follows: "Mathematics is precisely the ideal handling of the problems of life, and the central ideas of the science, the great concepts about which its stately doctrines have been built up, are precisely the chief ideas with which life must always deal, and which, as it tumbles and rolls about them through time and space, give it its interests and problems. . ." (p. 77).

What does he identify in a more fine-grained way as the connecting links? Keyser comments,

> The mathematical concept of constant and variable are represented familiarly in life by the notions of fixedness and change. The concept of equation or that of an equational system, imposing restriction upon variability, is matched in life by the concept of natural and spiritual law, giving order to what were else chaotic change and providing partial freedom in lieu of none at all. What is known in mathematics under the name of limit is everywhere present in life in the guise of some ideal, some excellence high-dwelling among the rocks, an 'ever flying perfect' as Emerson calls it. . . . The supreme concept of functionality finds its correlate in life in the all-pervasive sense of interdependence of mutual determination among the elements of the world. What is known in mathematics as transformation . . . is conceived in life as a process of transmutation by which, in the flux of the world, the content of the present has come out of the past and in its turn, in ceasing to be, gives birth to its successor. (p. 78)

Thus we arrive at an even more robust view of what it might mean to apply mathematics to the real world. It is already "in" the real world, and it is only by making believe that it is severed from the world that we arrive at some artificial notion of "application." That is, the application is there by virtue of the very way we speak about events in the world. The situation is analogous to the use of Freudian language in expressing everyday occurrences. It is not that we need to explain the meaning of Freudian language by recourse to ordinary language usage, but rather that a Freudian perspective has so infiltrated our everyday language that we cannot help but see the (supposedly non-Freudian) world through Freudian lenses.

Morality: A Principled Perspective
Religious education has made use of thinking in mathematics (as well as in other secular fields) for a long time. La Noue (1962) has shown, for

example, that problems of the following sort have been used in religious school texts in order to indoctrinate religious values in a subtle way. "The children of St. Francis School ransomed 125 pagan babies last year. This year they hope to increase their number by 20%. If they succeed how many babies will they ransom this year?" (p. 272). Aside from the problem that this example is offered in the service of religious rather than secular education, it is used, not as an invitation to *discuss* moral or ethical values or principles, but rather to *inculcate* values in bypassing such analysis.

The example of the comatose friend introduced earlier in this chapter says something about how ethical issues may be embedded in a problem that has mathematical components as well. It also invites decision making of an ethical sort and is thus a level removed from the St. Francis example above. There is much more that we can do, however, to encourage an explicit conversation on the nature of ethics in relation to mathematical thinking. What might we search for if we wish to make use of the scheme of Figure 5.5 in such a way that the mathematical examples may enable us to bring the nature of ethical thinking closer to the surface?

Without embarking on a discourse of the nature of ethics in general, there are some aspects of ethical thought and behavior that we might highlight for the purpose of trying to understand connections between ethics and mathematical experiences. It is important to appreciate, however, that we are not attempting to inculcate specific moral behavior, but rather are seeking ethical principles that can become part of the conversation as the result of mathematical exploration.

Clearly cultures and subcultures have different ways of behaving, and many of them adopt specific moral values that may be in conflict with each other. Nevertheless, there are certain *principles* many ethicists agree we would have to adopt if we were to claim that a way of behaving belonged to the category of morality.

One such principle is a principle of impartiality. It asserts that in order to behave in a moral way, we cannot justify our own behavior as moral and that of someone else as immoral if the other person were in a position that is essentially the same as ours.

That is to act morally is to act in such a way that we would accord anyone in a similar (should we say 'identical'—an issue that is raised by the discussion of "same/different" in the previous chapter) situation the same right to behave as we do. Of course we could drive a Mack truck between the crevices of what constitutes behavior that is "essentially the same" for two different people in supposedly similar circumstances. Nevertheless, the principle has merit as a heuristic for encouraging us to justify why we might behave in ways that are different from that of others who appear to be in similar circumstances.

We all surely behave at times in ways that do not generate pride. We all behave in ways that we would might not be able to justify as moral behavior. However, what we could not do if we take the concept of morality seriously is to justify our own behavior as morally right because, for example, "it makes us feel good," while at the same time to condemn the behavior of someone else in the same situation, implying that they do not deserve the same privileges. Young children are especially good at invoking the principle of impartiality when they ask why they can or cannot behave in certain ways when in fact their parents set an example of behaving otherwise. Frequently, the parent invokes the category of age. That is, because they are older, they tell their children that they are justified in behaving differently. The critical question from the point of view of morality, however, is whether or not age in fact constitutes a variable that makes the circumstances significantly different. Sometimes, it does, but not always, and it is the intuition that "not always" is a viable category that gnaws away at children and prevents them from being badgered into submission.

In adopting the principle of impartiality, we need to strenuously seek out how it is that the circumstances in one situation may be different from those in another. We might even want to make a point of honoring what is unique and unusual about another person's behavior and circumstances rather than jumping quickly onto an impartiality bandwagon primarily because it diminishes our need to rethink what and how we ought both act and react to another's behavior (or our own).

Part of moral education involves an invitation to evaluate the principle of impartiality itself and to think through its practical limitations, its theoretical presuppositions, as well as its potential to be in conflict with other humanly important ways of acting in the world that may not bear directly on morality.

How might we enable students to both appreciate some variation of the above principle and to feel what it is like to act in accord with it? At the very least it would seem that one needs considerable opportunity to imagine being in someone else's shoes. Specifically, we need to encourage ways of thinking that provide an antidote to egocentric thought.

The following two mathematical examples invite precisely such behavior. They do not necessarily *inform* students of what it feels like to be in someone else's shoes, but they do require that the students place themselves there.

Consider the contrived but intriguing problems:

(1) First Problem:
A debonair census taker in his early forties, making the rounds for the day, rings the doorbell of a home on a tree-lined street in a friendly neighborhood at 9:15 A.M. There is no answer at first, but after a third ring, he hears someone rushing down a flight of stairs. An attractive woman, whose age could be anywhere from early forties to mid fifties, dressed in a lightly clad negligee, answers the door. Here is the conversation:

Census Taker: Hello, I am the census taker.
Woman: Oh, have you been doing that for long?
Census Taker: No, not really, I am a math teacher, but I am doing this as a summer job.
Woman: I used to teach math also, but I am now in the insurance business.
Census Taker: Interesting . . . Where did you teach?
Woman: I used to teach at the high school around the corner.
Census Taker: Why did you leave?
Woman: Well, I liked a lot of the teaching, but one part of the curriculum that was central drove me bonkers, and it was something I could not learn to teach well.
Census Taker: What's that?
Woman: Word problems. Some kids loved them, and I barely had to teach anything for them to learn to handle those problems. But the majority hated the problems and tried to do them in mechanical ways rather than thinking things through. I went to loads of conferences and spoke to lots of colleagues, but I felt I could do little more than make my students feel inadequate when I told them not to treat the problems mechanically, but try to think through them.
Census Taker: I know what you mean. They were intended originally, not only to show kids how math relates to the real world, but to encourage them to think in more creative ways, and for the most part the problems are contrived, of little interest, and students try to handle them by a quick category scheme and then using some formula.
Woman: And for the most part the kids can see through the problems even if they cannot solve them. They realize that

Connecting Mathematics With the Real World

	the story line is weak and only a veneer that is presented for the purpose of getting on with the math.
Census Taker:	That's probably one of the reasons that the new reform movement has tried to eliminate the sort of story problems you used to teach, and instead replaced them with "real-world" problems.
Woman:	Come in for a cup of coffee and you can tell me about it. Maybe I should think of going back to the high school.
Census Taker:	O.K. It would be great for you to return to the classroom. You look like you would be a terrific teacher. Can we take care of a little census business before discussing this further?
Woman:	Sure, what would you like to know?
Census Taker:	Well, how many of you live here?
Woman:	There are four of us . . . my three daughters and myself.
Census Taker:	Oh? so you are not married?
Woman:	No, I have been happily divorced for almost six months now.
Census Taker:	I won't ask your age, though you look like you are in your mid-twenties, but how old are your three daughters?
Woman:	Let's see. If it were anyone else, I would answer that question directly, but appreciating what we have in common, let me put it this way: The product of my 3 daughters' ages equals 72. And the sum of their ages equals the address which you have already observed on the house.
Census Taker:	That's the kind of problem I love. Do you mind if I sit down and try to work it out with a pencil and paper?
Woman:	Be my guest.
Census Taker:	[Works things out for about ten minutes]. You know what? I do *not* have enough information to figure it out.
Woman:	You're right. I forgot to tell you that my oldest daughter has a cat with a wooden leg.
Census Taker:	Great! I can figure it out now. Thanks, that was a clever way to tell me.
Woman:	Would you like some buns with the coffee?
Census Taker:	I would love them, but I must leave now to catch up with my roommate who is also a math teacher. He leaves this afternoon to give a talk at a conference for teachers about

	the relationships between math and morality. This problem would be right up his alley.
Woman:	Which problem?
Census Taker:	The one you just gave me about the product of your daughters' ages.
Woman:	[comments absent-mindedly] Oh, that problem . . .

Based upon the information in the dialogue, can you figure out the ages of the daughters?[7]

(2) Second Problem:
Three women are sitting in a circle facing each other. A fourth one tells them that either one, two, or all three of them have a dark spot on their foreheads. Except for observing each other, they are not permitted to communicate with each other. Anyone who can prove that she has a spot in the middle of her head is to raise her hand and she wins a prize. Assuming that all three do in fact have spots in the middle of their heads (but none of the three is told that such is the case), can anyone establish that she has a spot on her forehead?

Consider the first problem. It looks at first as if you have two significant bits of information:

(a) the product of the ages is 72.
(b) the sum of their ages is the street address

We normally associate mathematical thinking with an effort to strip the particularity of a situation and to attempt to deal with abstractions. So, for example, we do not think much about the fact that it is a woman who answers the door or that she is talking about daughters rather than sons, or that she is at her house, and so forth.

Given this general orientation, in trying to make sense of this problem, it took a while to understand the significance of the fact that not only is the street address the sum of their ages, but

(c) the census taker actually gazed upon the information conveyed in (b).

As importantly,

(d) he had found that information compelling, but
(e) something was missing until he found out something about her oldest daughter.

Connecting Mathematics With the Real World 159

In order to see the significance of the given information, we cannot place ourselves totally outside the scene. We almost have to picture ourselves *in* the situation. We have to be careful not to abstract too much. Before pretending that you are the person actually taking the census, it seems perfectly reasonable to accept the conclusion that you do not have enough information.

In fact, I gave this problem to several of my students recently and those who had a more intense and specialized mathematics background were most inclined to conclude that the problem was impossible to solve. They had begun by setting up some equations and concluded based upon the number of unknowns and the number of equations that there was an inadequate amount of information to solve the problem.

After actually imagining yourself in the scene (as the census taker who observes what the woman is pointing at) and after you have thought about what the possibilities might be (as an outsider), then it seems that you probably do have enough information. As a person in the scene you imagine that you would know the answer to the census question—although an outsider (the person doing the problem) would not know it.

The leap that led me to the solution was the realization that I needed to find a way of seeing the problem as an outsider so that I could reexamine the circumstances that might have led the census taker to conclude that without knowing the last bit of information about the oldest daughter's cat, the problem was too ambiguous.

Putting yourself in someone else's place, stepping out of the scene, shifting your expectations dramatically, being cautious about what you abstract and what you view as particularity, imagining alternatives, relying upon and revising intuitions, coming up with an anti-intuitive realization, (one of the last steps for me), juxtaposing inductive and deductive thought, are all "real-world" experiences that are rarely captured in more than a superficial way when connectedness is limited to the concept of model we discussed earlier in this chapter.

With the possibility that the reader may be grabbed by this problem (taking into consideration our earlier caveat regarding what "hooked" might mean), and not wanting to spoil too much should that be the case, let me suggest some possible triples for their ages:

$$2, 3, 12$$
$$2, 4, 9$$
$$3, 4, 6$$

In each case the product is 72. Take a look at the sums, the possible street addresses. They are 17, 15, and 13, respectively. I have not listed all the possibilities, but if further exploration reveals that the sums are all different (as they are so far), then we would conclude that the census taker would have been able to figure out the correct triple by knowing the precise sum—even though we as outsiders would not know the answer.

Now, what is there that might prevent the census taker from knowing the correct answer once he knows the sum to be the same as the street address? One scenario that could make it impossible for him to know the ages if he is given only the sum, would be if there are two different sets of triples that yield the same sum. Finally, what could conceivably be relevant about reference to the oldest daughter? (See note 6 for a hint if you are climbing the walls).

Before turning to the next problem, it is worth seeing where we have been headed. What do we conclude from the fact that a solution requires some fascinating interplay of stepping in and out of the scene—or from placing oneself in someone else's position and stepping back into some more neutral observer's position?

Though experiencing the world from someone else's perspective in a mathematical setting cannot be *harmful* with regard to acquiring moral sensitivity, the mathematical experience in and of itself will not necessarily encourage students to come to terms with notions of morality. This sort of activity is valuable not for what it accomplishes inadvertently, but for the invitation it offers to compare moral and mathematical experiences.

We need to encourage students to ask questions such as: What is there about placing yourself in someone else's position in this problem that is like what is done in the moral domain? What is different? What more may be needed if one is to think and behave morally?

These types of problems provide the impetus for a new kind of conversation. More precisely, what is learned through the act of putting yourself in someone else's place? What is there about such activity that is above and beyond what we have done in this example? In particular, what are the cognitive, emotional, and epistemological dimensions of empathy itself? (See Verducci, 2000 for an excellent summary of the different conceptions.)

Without going into much detail in the case of the second problem, let me suggest that you imagine that you are one of the three characters looking at the other two women. This is a first level of approximation of putting yourself in someone else's position. More thought is needed here, for as you look out at the other two women, you notice that (a) each has a spot on her forehead, but (b) neither of them raises her hand to claim that she believes that to be the case. In order to make some headway here, it is now necessary to think through what must be in the minds of the other two women if they are not raising their hands. A good beginning might be to imagine what would have enabled them to in fact raise their hands.[8]

Morality and mathematics share many other domains.[9] Dealing with their commonalties and differences has the potential to further enrich how we view each of the fields. We can, for example, borrow some ideas from our discussion in the previous chapter on same/different, as well as on wonder. It turns out that our intuitions and more rigorous analyses of issues in each of these areas occasionally come in conflict with each other. Just as it is shortsighted to adopt the advice that from a mathematical point of view rigorous analysis automatically takes precedence over intuition, so it is a mistake to rank order those ways of thinking in morality as well. Frequently we may be able to sense without being able to articulate some critical issues that are relevant in terms of making moral decisions. The situation we encounter in each of the fields may be so unique that we are not able to see with any degree of rigor what it is we need to assume in order to operate wisely.

The history of mathematics is studded with such examples, especially in relation to infinity as we described earlier. If we track the history of the problem of determining the sum of the following infinite progression, we find all sorts of supposedly rigorous analyses that do not necessarily enable us to decide the issue. Consider, for example, the question of what the value of the following infinite sum should be:

$$1 - 1 + 1 - 1 + 1 - 1 + 1 - 1 + \ldots = ?$$

In order to appreciate that merely clarifying the problem by invoking reasonable notation is not adequate to reach a decision, consider which of the following is a more reasonable way of choosing an answer:

$$(1 - 1) + (1 - 1) + (1 - 1) + (1 - 1) + \ldots = 0$$
$$1 + (-1 + 1) + (-1 + 1) + (-1 + 1) + (-1 +) + \ldots = 1$$

How should we look at the two possible ways of finding the infinite sum? There were some mathematicians, at the time this issue emerged and was hotly debated, who suggested that since both seemed reasonable, we should compromise and choose 1/2 as the answer. This problem had challenged mathematical genius for almost a century. To arrive at an acceptable twentieth-century answer, an agreed upon notion of what it means to determine the sum of an infinite number of terms was needed, as well as a clearer understanding of the established principles for finding the sum of a finite number of terms. There was nothing god-given about the resolution, and surely we could come up with some scheme that challenges the supposedly rigorous conclusion that—because an associated sequence of numbers does not have a limit—there is no number at all that qualifies as the infinite sum.

Morality: A Feminist Perspective

Before turning to the next application of mathematics to the real world, we highlight another possible connection between mathematics and morality. That is, the relationship between ethics and morality to which we alluded at the beginning of this section is a tricky one. We spoke of the principle of impartiality, which is "empty" in the sense that it says nothing about specific behavior but rather about what would be expected of any system that deals with morality. There is surely an abstract quality to this principle and to such conceptions of morality.

Some of our discussion, above hinted at, and in fact invoked, an element of an alternative view of morality. The fact that we framed the issue of impartiality against the backdrop of "putting yourself in someone else's shoes," suggests that we were straddling a dispassionate principled perspective and another point of view that has influenced education significantly. This alternative viewpoint has been developed by psychologists and philosophers of education beginning in the early 1980s.[10] It can be seen as a reaction to a popular view of moral education that was developed by Lawrence Kohlberg. Much of Kohlberg's work was driven by the use of moral dilemmas as a form of moral education of youngsters. We now place that work in perspective.

A typical dilemma he has used for much of his research and for his deliberate program of education as well is the Heinz dilemma, below:

> In Europe, a woman was near death from a rare form of cancer. There was one drug that the doctors thought might save her, a form of radium that a druggist in the same town had recently discovered. The druggist was charging $2000, ten times what the drug cost him to make. The sick woman's husband Heinz, went to everyone he knew to borrow the money, but he could only get together about half of what the drug cost. He told the druggist that his wife was dying and asked him to sell it cheaper or let him pay later. But the druggist said, 'no.' So Heinz got desperate and broke into the man's store to steal the drug for his wife. (Kohlberg 1976, p. 42)

Should Heinz have stolen the drug? Based upon an analysis of longitudinal case studies to answers of dilemmas of this sort, Kohlberg has created a scheme of moral growth that he claims is developmental. Furthermore, he has created not only a research tool but an educational program around such dilemmas. It is through discussing and justifying responses to such dilemmas that students mature in their ability to find good reasons for their choices. It is not the specific value that one chooses (e.g., steal the drug versus allow the wife to die), but the reasons offered for the decision that places people along a scale of moral development.

At the lowest level of moral maturity (preconventional) Kohlberg finds that people argue primarily from an awareness of punishment and reward. Thus someone at such a stage of development might claim that Heinz should not steal the drug because he would be punished or sent to jail, or he might claim that he should steal it because his wife might pay him well for doing so. It is almost as if the punishment inheres in the action itself. At a later stage (conventional) people might argue from the more abstract perspective of what is expected of you and also from the point of view of the need to maintain law and order. At the highest stage of principled morality, according to Kohlberg, one argues on the basis not of rules that could conceivably change but with regard for abstract principles of justice and respect for the dignity of human beings. Such principles single out fairness and impartiality as part of the very definition of morality. None of these structural arguments (e.g., punishment/reward, law and order, justice) in themselves dictates a correct resolution of any dilemma. Rather, they form part of the web that is used to justify the decisions made, and it is by listening to these reasons that Kohlberg and his followers are capable of deciding upon one's level of moral maturity.

Despite the fact that Kohlberg's scheme for negotiating moral development neglects to focus upon action, it is a refreshing counterpoint to a program of moral education which conceives of its role as one of inculcating specific values in the absence of reason. Nevertheless, there has been some penetrating criticism of his scheme—a criticism condemning Kohlberg's work on the grounds of sexism. Specifically, Kohlberg's research and ultimately his scheme for what represents a correct hierarchy of development is based upon his longitudinal research *only with males*. Once the scheme was created and the stages developmentally construed, Kohlberg interviewed females and concluded that their deviation from the established hierarchical scheme implied an arrested form of moral development. Carol Gilligan (1982) points out that the existence of a totally different category scheme for men and women not only may be a consequence of different psychological dynamics, but rather than exhibiting a logically inferior mind set, women's handling of these dilemmas, suggests that Kohlberg's moral categories are in need of reconstruction.

Compare the following two responses to the Heinz dilemma, one by Jake, an eleven-year-old boy and the second by Amy, an eleven-year-old girl. Jake is clear that Heinz should steal the drug at the outset, and justifies his choice as follows:

> For one thing a human life is worth more than money, and if the druggist makes only $1000, he is still going to live, but if Heinz doesn't steal the drug, his wife is going to die. (Why is life worth more than money?) Because the druggist can get a thousand dollars later from rich people with cancer, but Heinz can't get his wife again. (Why not?) Because people are all different and so you couldn't get Heinz's wife again. (Gilligan, 1982, p. 26)

Amy on the other hand equivocates in responding to whether or not Heinz should steal the drug:

> Well, I don't think so. I think there might be other ways besides stealing it, like if he could borrow the money or make a loan or something, but he really shouldn't steal the drug—but his wife shouldn't die either. If he stole the drug, he might save his wife then, but if he did, he might have to go to jail, and then his wife might get sicker again, and he couldn't get more of the drug, and it might not be good. So, they should really just talk it out and find some other way to make the money. (Gilligan, 1982, p. 28)

Notice that Jake accepts the dilemma and begins to argue over the relationship of property to life. Amy, on the other hand, is less interested in property and focuses more on the interpersonal dynamics among the characters. More importantly, Amy refuses to accept the dilemma as it is stated, but is searching for some resolution that invokes less of a zero sum game mentality. Kohlberg's interpretation of such a response would imply that Amy does not have a mature understanding of the nature of the moral issue involved because she fails to appreciate that this hypothetical case is designed to test whether the subject appreciates that in a moral scheme life takes precedence over property. Gilligan, on the other hand, in analyzing a large number of such responses has concluded that the females are not arrested in their ability to move through Kohlberg's developmental scheme. Rather, they tend to abide by a system which is orthogonal to that developed by Kohlberg. In this system the concepts of *caring* and *responsibility* rather than *justice* and *rights* ripen over time.

Behind the female voice of responsibility and caring, some of the following characteristics surface:

(a) a context-boundedness;
(b) a disinclination to set general principles to be used in future cases;
(c) a concern with connectedness among people.

Although not all of these characteristics are exhibited in Amy's response, they do appear in interviews with mature women (Gilligan, 1982; Lyons, 1983). Context-boundedness represents a plea for more information that takes the form not only of requesting more details (e.g., the relationship between husband and wife) but of searching for a way of locating the episode within a broader context. Thus unlike men, mature women might respond not by trying to resolve the dilemma but by exhibiting a sense of indignation that such a situation as the Heinz dilemma might arise in the first place. They might say, "The question you should be asking me is 'What are the horrendous circumstances that caused our society to evolve in such a way that dilemmas of this sort could even arise—that people have resorted to such acts of miscommunication'"? The second characteristic is an attempt to understand each situation in a fresh light, rather than in a legalistic way, i.e., in terms of established precedent. Connected with context-boundedness is the desire to see the fullness of the situation and to recognize how it might be different from (and thus require fresh insight) rather than be easily derived from a situation that has already been settled. With regard to the third characteristic, conflict is less a logical puzzle to be resolved but rather an indication of an unfortunate fracture in human relationships, i.e., something to be "mended" rather than an invitation for some judgment.

Much of what we have criticized and reformulated regarding both problem solving (in chapters 2 to 4) and connections with the real world (in this chapter) resonates well with a feminist perspective on morality. The distinction between *problem* and *situation*, for example, calls for a personal connection with the subject matter. By asking what it is that makes a situation problematic, it is necessary to dig deeply into what we perceive to be concomitants of an apparently isolated phenomenon. More generally, the focus on problem posing, as opposed to problem solving (especially use of What-If-Not ways of thinking) invites the student not only to create context, but as importantly to challenge rather than to accept what is given as a logical puzzle.

Furthermore, by viewing the concept of model differently, we come to appreciate the potential particularity of a situation or problem rather than seeking to locate what is easily dismissed as "irrelevant." In searching for more rather than less information, we transform the nature of our inquiry significantly. It is in this spirit that questions we discussed with regard to problems take on added significance.

By asking questions of the following sort, we are acknowledging a feminist perspective. In fact, it is inherent in much of what we criticized with regard to problem solving and model as the connection between mathematics and the real world.

1. What purpose(s) are served by my solving this problem or this set of problems?
2. Why am I being asked to engage in this activity at this time?
3. What am I finding out about myself and others as a result of participating in this task? [11]
4. If I find difficulty understanding this problem (situation), what might be the reasons?
5. How might I use this problem (situation) to create others that may or may not share essential qualities with what I am "given."
6. What light does this particular problem (situation) shed on what I think of as mathematical inquiry?
7. How is the relationship of mathematics to society and culture illuminated by my studying how I or other people in the history of the discipline have viewed this phenomenon?
8. Do I care about this problem (situation)? What would it take for me to care differently about it?

"Black" Humor

How might mathematical thinking be related to humor? This question itself might seem a tad funny (peculiar?). For most people, the association of mathematics with humor is absurd. To seek connections, it will be helpful to explore what it is that theorists of humor have to say about the nature of humor.

In this regard, one of the cleverest lectures I have ever heard was given in the late 1970s by Max Black, a Cornell philosopher (now deceased), who had become interested in the philosophical analysis of humor late in life. The lecture began with a variation of the following two jokes:

1) Mr. Smith, an American in Paris who knew very little French, was eating in an exclusive French restaurant. He noticed that there was a fly in his soup. he called the waiter over, took out his French-English dictionary and pointed at the fly in the soup.

Mr. Smith: Garçon, *un mouche, un mouche.*
Garçon: Non monsieur, *une mouche, une mouche.*
Mr. Smith: What fantastic eyesight!

2) Mr. Gould meets an old friend, Mr. Brown, in the street. They had not seen each other for many years. The following conversation ensues:

Mr. Gould: So, how are you?
Mr. Brown: Awful.
Mr. Gould: Why?
Mr. Brown: I just came back from the doctor with my wife and found out that she has an incurable disease and has only six months to live.
Mr. Gould: It could be worse.
Mr. Brown: And also I was just in an automobile accident and my brand new car was demolished and I had no collision insurance.
Mr. Gould: It could be worse.
Mr. Brown: I got a phone call last week from my son, who has been married for twenty five years and he tells me that his wife, whom I adore, and who has been wonderful to me and my wife, is planning to divorce him.
Mr. Gould: It could be worse.
Mr. Brown: My partner came to my house yesterday and told me that he has been cheating our customers for years. Someone has informed the government and we are being sued.
Mr. Gould: It could be worse.
Mr. Brown: What do you mean by telling me "It could be worse?" I tell you my wife is dying; my son is about to get a divorce; my car is destroyed, and I am about to become bankrupt. What could be worse?
Mr. Gould: It could be me!

It is not so much that the jokes were hilarious (though they were clever). Rather it was the intent of the jokes that made the lecture so clever. Max Black continued with about a dozen additional jokes. He then challenged the audience to come up with a single theory of humor that covered all of the jokes.

The Nature of Humor
Mindful that joke telling and humor are not synonymous, Black was using this format to explore the question of whether or not humor can be reduced

to one theory. His point was that it was not possible to do so. There are, in fact, a number of different theories of humor, and it remains an interesting question whether or not they can be consolidated under one theory.[12]

This is a question that is explored by Paulos (1982) in an insightful work entitled, *Mathematics and Humor*. He reviews some of the major theories of humor and does in fact come up with a consolidated theory. Based upon that theory, he finds some fascinating connections between mathematics and humor. I will briefly review the theories; then his consolidation and proposed connection with mathematics; finally I will propose an additional connection that he has left out.

One of the earliest theories of humor is due to Aristotle, who wrote a lot more about tragedy than humor. The Greek conception of humor associates it with base instincts. It has to do with what is ugly but not disastrous. Derived in part from the above is a theory of humor that was popularized by Thomas Hobbes (1914) in the *Leviathan*. His theory, known as "sudden glory" associates humor with a sense of superiority, a realization that one's station is superior to that of the person(s) who are seen as humorous.

Freud (1960), not surprisingly, had a theory of humor that spoke more about its function than its qualities. In his *Jokes and Their Relation to the Unconscious*, he claims that humor enables people to vent their aggressive and repressed sexual feelings.

The best known theories of humor are those associated with Henri Bergson, Arthur Koestler, Molière, and others. There are many variations of this perspective, but the connecting feature is that humor is associated with incongruity. Paulos adopts a variation of this theory in seeking connections with mathematics. He points out correctly that if incongruity is a necessary condition for humor, it surely is not sufficient, for there are many incongruous events, ideas, circumstances that are not at all funny. Paulos claims that in addition to *having* incongruity, humor requires (1) that the incongruity be noticed; (2) that it have a point; (3) that the emotional climate must be right.

Humor as Ingenuity: Connection With Mathematics

Where does mathematics enter the scene? For Paulos it begins with the realization that ingenuity and cleverness are hallmarks of both mathematics and humor. Of course, he is not referring to use of mindless techniques or overused algorithms, but rather to unusual or unexpected strategies.

Connecting Mathematics With the Real World 169

Some of what we discussed in chapter 2 (mathematical genius) and chapter 4 (on wonder) exemplifies the sort of cleverness he has in mind. The ability of Gauss to find the sum of the numbers from 1 to 100 without resorting to algorithmic calculation surely has a cleverness to it that is akin to what we find in a great deal of humor. It is an example of incongruity in the sense that the clever point of view appears to be coming from "out of left field." Once it is produced, we may be able to appreciate how it fits, but such cleverness is not apparent as we begin to think about the problem. In brief, it may be easy to verify but difficult to produce.

There is a deeper and more easily recognized connection that Paulos locates in the relationship between concrete and abstract axiom systems (discussed in this book earlier in different contexts). It is a very keen observation on his part and in fact helps us to understand the nature of at least one form of humor.

Humor and Abstraction

Paulos recalls the following classical burlesque joke as an example of the connection (somewhat sexist) between two different levels of abstraction: "The dirty old man leers at the innocent young virgin and says, 'What goes in hard and dry and comes out soft and wet?' The girl blushes and stammers, 'Well let's see, uh . . . ,' to which the dirty old man replies wickedly, 'chewing gum.'"(p. 24).

To appreciate the above joke, it is necessary to see it as akin to the confusion of an abstract system with its concrete realization. As we depicted in chapter 1 with regard to the student who was learning vector spaces as an abstract structure, such structures are in some deep sense devoid of meaning. Abstract structures speak of elements and relations but do not specify what the actual elements and relations are.

The power of such abstract characterizations is that they have application to many concrete realizations. If we come to see that two concrete systems are structurally the same, then it is no longer necessary to reestablish truths in one system if we have already done so in another.

Let us return to our discussion of isomorphic structures with a different focus. What makes the chewing gum joke funny is that the young woman is holding one model in mind while the dirty old man has another. The latter is a model that is understood to be structurally similar to the former, but it is not anticipated until after it is produced. We then can see (intuitively) that they both share the same abstract structure. The incongruity is not that they are different, but that the two systems (which from some important points of view are different—especially with regard to a sexual charge) are basically the same.

How and when one is taught about abstract systems in mathematics is of course a critical question. Though *The Standards* were motivated in large part by a realization that the "new math" had gone overboard in this direction, it is surely a perspective that one cannot dissociate from mathematics education. How we come to appreciate an axiomatic perspective and how we choose one system over another is surely an important mathematical activity. If we do not view axiomatics as born like Athena, fully grown from Zeus' head, and if we appreciate the sense in which any system is both socially constructed and changeable over time, then there surely is a need to include that perspective as students mature in their mathematics education.

Humor and Self-Referentiality

Yet another perspective on the relation of humor to mathematics has to do with the theme of self-referentiality. The concept of self-referentiality is evoked, loosely speaking, whenever an object refers to or calls upon itself. I was first aware of experiencing the concept explicitly a number of years ago when my daughter Sharon, who was four years old asked me, "If God created everything, then who created God? You tell me that!" It is of course a beautiful theological question and I will not pretend to be able to offer a response. Though some might dismiss rather than answer the question, others might assume that the question is both meaningful and difficult to answer. One of the delightful qualities of the concept of self-referentiality is that it frequently appears to be on the cusp of meaninglessness. Groucho Marx who is reputed to have exhibited one of the most famous self-referential comments when he made the whimsical remark, "I would never join a club that would accept me as a member." Similarly Tom Lehrer et al. (2000), a well known satirist in the 1950s to the 1970s remarked as a prelude to one of his songs, "I know there are people in the world who do not love their fellow human beings, and I hate people like that."

Not only verbal jokes, but visual images frequently evoke a self-referential quality. In *Mathematics and Humor*, Paulos (1982) has a cartoon of people sitting in a train beneath an advertisement which reads, "Want to Learn to Read?" It then gives a telephone number for further information. The logic of the cartoon is "grabby." Anyone who knows how to read does not need to invoke the services of the organization. On the other

Connecting Mathematics With the Real World 171

hand, anyone who does not know how to read, has no way of knowing what is being proposed by the advertisement in the first place. Paulos refers to this cartoon as an example of self-contradictory self-reference. In instances like this, the content and form or mode of expression are incongruous.

Douglas Hofstadter in his Pulitzer prize winning work, *Gödel, Escher, Bach: An Eternal Golden Braid*, shows how self-referentiality connects the fields of art, music, and mathematics. Musical use of self-referentiality is expressed in Bach's magnificent fugues—a less formalized notion than a canon. Hofstadter expresses the self-referential quality of a canon as follows:

> The idea of a canon is that one single theme is played against itself. This is done by having "copies" of the theme played by the various participating voices. But there are many ways to do this. The most straightforward of all canons is the round, such as "Three Blind Mice," "Row, Row, Row Your Boat," or "Frère Jacques." Here, the theme enters in the first voice and, after a fixed time-delay, a "copy" of it enters, in precisely the same key. After the same fixed time-delay in the second voice, the third voice enters carrying the theme, and so on. Most themes will not harmonize with themselves in this way. In order for a theme to work as a canon theme, each of its notes must be able to serve in a dual (or triple, or quadruple) role: it must firstly be part of a melody, and secondly it must be part of a harmonization of the same melody. When there are three canonical voices, for instance, each note of the theme must act in two distinct harmonic ways, as well as melodically. Thus, each note in a canon has more than one musical meaning; the listener's ear and brain automatically figure out the appropriate meaning, by referring to context. (p. 8)

The application of self-referentiality is exemplified in art as well. It is one of several cleverly used schemes by the Dutch artist, M. C. Escher. Although some of his self-referential art is quite elaborate, a rather simple and elegant one involves a sketch of two hands ostensibly facing each other, almost in shaking position. At the end of each hand is a pen which is drawing something. What would you guess it is drawing? One hand is, in fact, drawing the other, but there is no way to decide which hand is actually drawing which.

How does mathematics participate in this adventure? It turns out that the concept of self-referentiality or its near relatives flourish in quite diverse fields, from mathematical logic to number theory, from chaos theory to combinatorics, from geometry to computer science as well.

A simple example from logic itself would be:

i "This sentence is between five and ten words long."
ii "This sentince has three erors."
iii "This sentence is false"

Though it refers to itself, sentence (i) above is both straightforward and true. Sentence (ii), however, is delightful and not only exemplifies self-referentiality, but itself verges on the humorous. It looks as if there are only two errors in the sentence, both spelling errors. So what is the third error? Look again at the sentence. The third error talks not about the individual constituents of the sentence, but of the sentence as a whole. That is, the claim is made that the sentence has three errors, but that claim about the sentence itself is wrong, for it has only two errors. Therefore, the third error in (ii) above is the claim that the sentence has three errors!

Sentence (iii), however, is more interesting. Its humor resides in its playfulness, a kind of playfulness that is depicted in the cartoon of Paulos that we described above. Actually this kind of analysis had a particularly significant impact on the foundations of mathematics. In mathematics sets were taken to be powerful, fundamental, and unproblematic building blocks from the nineteenth through the early part of the twentieth century. Bertrand Russell (1908) brought the theory to a screeching halt by showing that it was not so easy to speak of sets as a clearly defined concept. He essentially created an innocently appearing new set out of two old ones, which led to some quite problematic consequences.

Russell argued that there are essentially two types of sets. Specifically, some sets are members of themselves and some are not. For example, the set consisting of all ideas in the universe is itself an idea. Such a set is therefore a member of itself. The set consisting of all those sets that are meaningful is itself meaningful. This set also is therefore a member of itself. The set consisting of all irrational numbers is not itself an irrational number, however, and is therefore not a member of itself. Likewise the set consisting of all sets with less than five members has more than five members (in fact an infinite number) and therefore is also not a member of itself.

Now Russell created a new and curious set: the set of all sets that are not members of themselves. Let us call that set R (for Russell). He asked of this set an innocent sounding self-referential question: Is set R a member of itself?

If you think about that question in the privacy of your own boudoir, and if the light is shining at the right angle, and if you are in an upbeat mood, you will probably come to something like the following conclusion:

If R is a member of itself, then since R consists only of those sets that are not members of themselves, R cannot be a member of itself. On the other hand, if R is not a member of itself, then R must be a member of itself since the set R captures only those sets that are not members of themselves.

With this exposé, mathematicians and philosophers of mathematics could no longer feel secure with selecting sets as building blocks for the rest of the discipline. Russell in fact created a concept called the theory of types in which he restricted the kinds of sets that it was possible to speak without fear of contradiction of the sort we discussed above.[13]

From the point of view of foundations of mathematics, the concept of self-referentiality goes a long way. It was the use of this concept that enabled Gödel (discussed in chapter 1) in the mid-1940s to engage in something every bit as devastating as Russell's finding. The analysis crushed the mathematician's optimism that the field could be viewed as the last threat to the "Work Ethic." Through the use of self-referentiality on a scheme involving the natural numbers, he showed that intelligence and hard work have some inherent limitations in relation to proving of all mathematical truths.

We see then that mathematical self-referentiality shares much with humor (and other fields as we mentioned above). It is a delightful concept—a thing calling upon or talking about itself. The concept is explored not only in foundational fields, but in a variety of other subdisciplines as well, many of which are part of the standard secondary school curriculum, some part of the emerging curriculum, and even a part of the *stranded* (!) curriculum. It is not only that self-referentiality is expressed in both mathematics and humor, but also that the concept, whenever it is expressed, has a fanciful twist. It frequently mingles a sense of lightness and depth simultaneously.

One of the most recent curriculum concepts associated with self-referentiality is due to Benoit Mandelbrot. Born in Warsaw in 1924, he was a refugee from Nazi Germany, a dabbler in many applied fields and a self-made mathematician. Coining the word "fractal" in 1975 for unusual shapes he discovered in nature, he contributed significantly to the newly emerging field of chaos theory—seeking regularity in irregularity. Mandelbrot's (1999) theory applies to fields as diverse as geometry, coastlines, and the stock market.

Gleick (1987) describes some of Mandelbrot's fascinating search for regularity in reference to a question that Mandelbrot pursued about coastlines. The question was inspired after Mandelbrot realized that there was a great discrepancy among different well accepted sources on the length of coastlines of various countries. Gleick's description has a ring to it that educators will find endearing. He tells us what happened when Mandelbrot asked people questions like: "How can I determine the length of the coastline of England"?

> He found that most people answered the question in one of two ways: "I don't know, it's not my field," or "I don't know, but I'll look it up in the encyclopedia." In fact, he argued, any coastline is—in a sense—infinitely long. In another sense, the answer depends on the length of your ruler. Consider one plausible method of measuring. A surveyor takes a set of dividers, opens them to a length of one yard, and walks them along the coastline. The resulting number of yards is just an approximation of the true length, because the dividers skip over twists and turns smaller than one yard, but the surveyor writes the number down anyway. Then he sets the dividers to a smaller length—say, one foot—and repeats the process. He arrives at a somewhat greater length, because the dividers will capture more of the detail and it will take more than three one-foot steps to cover the distance previously covered by a one-yard step. He writes this new number down, sets the dividers at four inches, and starts again. This mental experiment, using imaginary dividers, is a way of quantifying the effect of observing an object from different distances, at different scales. An observer trying to estimate the length of England's coastline from a satellite will make a smaller guess than an observer trying to walk its coves and beaches, who will make a smaller guess in turn than a snail negotiating every pebble. (pp. 94–95)

Gleick points out that our common sense would lead us to believe that as we take these supposedly more detailed measurements, we will reach some upper limit that is the "true length" of the coast of England.

In fact this intuition is incorrect, though as Gleick says, it would be true that if the coastline were some Euclidean shape (an arc of a circle, for example). Then this method of adding smaller and smaller lengths of segments of straight lines would indeed converge to some limit. He then tells us, "But Mandelbrot found that as the scale of measurement becomes smaller, the measured length of a coastline rises without limit, bays and peninsulas revealing ever-smaller subbays and subpeninsulas—at least down to atomic scales, where the process does finally come to an end" (p. 95).

The advent of the computer facilitated exploration of complicated boundaries such as coastlines in ways that would have been virtually impossible earlier. Indeed, there are now computer programs accessible to youngsters that offer a glimpse at what it means for lengths that appear to be "visible" yet infinite.

One of the most famous of such examples is the fractal in Figure 5.7 known as the Koch curve.

Figure 5.7 The Koch curve created by continuing to replicate smaller equilateral triangles on the sides of an equilateral triangle.

Its self-referential quality is captured by the following description of the snowflake-like figure:

Take an equilateral triangle whose sides, let's say, are one inch long.

Take the middle third of each side and create other equilateral triangles on the outside. The length of each of their sides will be one-third the length of the sides of the original. We end up with a figure that looks like the star of David.

Now take each of the sides of the six equilateral triangles, and repeat the process, thus creating equilateral triangles on the middle portions of each of the sides. We now have eighteen little triangles as depicted in the third region in Figure 5.7.

If this process is continued indefinitely, we end up with a figure resembling a snowflake. Interestingly, it shares a property that is quite startling with Mandelbrot's coastlines. The length of the boundary of the figure—assuming that the process continues forever—is infinite. Nevertheless, the area of the snowflake is bounded. No matter how much it grows, it remains within the circle that is circumscribed by the original equilateral triangle.

Lest we lose sight of the (almost proverbial) snowflake for the larger picture of the dramatic winter scene, let us say a little bit about what we have been doing. That is, we have been exploring the concept of self-referentiality for the purpose of highlighting how it functions in a variety of contexts. While we have seen it displayed in many settings—including art and music—we were inspired originally to seek the connection between two quite unlikely bedfellows: mathematics and humor.

There is significant beauty, playfulness and surprise, in the concept of self-referentiality and its relatives. It has influenced literature and films in ways we have barely begun to explore. In chapter 1, we mentioned Woody

Allen's *Purple Rose of Cairo* (See Greenhut, 1985), and recalled that one of the characters portrayed in the audience at the beginning of the movie not only loses herself in the movie, but she becomes a character in its ongoing plot. It is becoming commonplace for modern novels to operate in such self-conscious ways. In Raymond Federman's *Smiles on Washington Square*, for example, the characters advance the plot by inventing each other. In other novels, characters not only decide to move the plot in a new direction, but they do so in conversation with the novelist.

The theme is a deep one and the more I think about personal matters, the more I find myself thinking in self-referential ways. It has become for me almost a metaphor for mind. That is, mind has the quality of asking itself "What is a mind?" Attempts at answering this question reveal unusual insights. Much of this book has already made and will continue to make use of self-referential ways of thinking even when bells are not sounded to tell the reader that we are so engaged. As mentioned above, one of the most interesting aspects of the concept is that it appears to be on the cusp of rationality. Sometimes it makes a lot of sense; sometimes it is confusing; sometimes it leads to surprising contradictions.

As a reminder that at least one popular variation of the concept is both central and problematic in mathematical thinking, consider the special form of deductive proof known as mathematical induction.

One version of its abstract formulation is as follows:

Let $P_1, P_2, P_3, \ldots$ be a sequence of propositions. Then if:

i. If P_1 is true;
ii. If P_n implies P_{n+1} for all n, then
all the propositions $P_1, P_2, P_3, \ldots$ are true.

A typical application of this form of proof might be to establish the validity of Gauss's discovery—the one we discussed in chapter 2. That is, suppose we want to show that:

$$1 + 2 + 3 + 4 + \ldots + n = [n \cdot (n + 1)] / 2$$

Without making use of Gauss's brilliant insight, we could do something a bit more "plodding" and make use of mathematical induction as follows:

i. P_1 is true. That is if we have the trivial case of just the first term with which to contend, then,

$1 = [1 \cdot (1 + 1)] / 2.$

ii. Assume P_n.
Then we need to show that this implies P_{n+1}.

We would then examine:
$(1 + 2 + 3 + 4 + \ldots + n) + (n + 1).$
Assuming P_n, we can re-express the above as
$[n(n + 1)] / 2 + (n + 1).$
However by the distributive property this is:
$(n + 1) \cdot [(n/2 + 1) /]$
This is then simplified as:
$(n + 1) \cdot [(n + 2)/2],$
which is of the form P_{n+1} since it is expressible as:
$[(n + 1) \cdot [(n + 2)]/2.$

Thus, the assumption that P_n is true does in fact lead us to conclude P_{n+1}.

Most people who have taught this form of proof have met interesting resistance. Even if students have no difficulty with the *technique* of establishing that P_n implies P_{n+1}, they initially feel that the reasoning is basically flawed. They believe that by assuming P_n as an essential step in the proof, they in fact are presupposing what they are trying to prove and therefore are engaged in circular reasoning. In brief, the process appears to be threatening rationality.

Clever use of metaphors (such as thinking about mathematical induction as being like a collection of standing dominoes with the understanding that knocking any one of them down has the effect of knocking the one next to it, and so forth), may change students' reluctance to engage in mathematical induction and may even be persuasive. A curriculum which takes a look at the panoply of ideas that employs self-referential thinking may not only persuade them of the justification of the enterprise of mathematical induction, but may also enable them to see that it is reasonable to have doubts.[14]

Mathematics and humor are linked in their need for and use of a variety of self-referential schemes. What is particularly intriguing is that no matter how rigorously the self-referential schemes are developed, we frequently find that they wrap the object of their affection in a cloth that is simultaneously deep and fanciful. A number of the mathematical examples we have reviewed in this section entice and beckon us because they appear to come with a tag that says, "Should I take this seriously or not?"

"Is someone just pulling my leg, or am I confronting an essential truth of the universe?"

I explore this double-edged facet further in the following subsection in a non-self-referential context.

Math *Is* Funny
Of all the connections between mathematics and humor, the most interesting is perhaps the simplest, though it has barely been explored. That is, from some points of view, mathematics is funny. What could that mean? There is an interesting and unusual sort of incongruity revealed in aspects of mathematical thinking. We turn now to that story.

We have already discussed a number of challenging problems—some solved, some unsolved—dealing with prime numbers. In chapter 2, we alluded to what was an intractable prime number problem in the set of natural numbers: the search for a formula that would generate prime numbers. Well known mathematicians, including Fermat and Mersenne in the seventeenth century, and Euler in the eighteenth century, devoted years of their lives to attempt to make headway on the problem.

Perhaps the most interesting way to understand the depth of the problem, and ultimately the humor involved, is for us to whittle away at the question being asked. If we know that the number of primes is infinite, then the simple formula "n" for all natural numbers n would do the trick. Well . . . yes, in a sense. The formula would generate all the primes and even do so in sequence, but at a price, i.e., it would generate all the nonprimes along the way as well.

The most ambitious request would be for a formula that generates all the primes in sequence and none of the composites. Short of that, we might hope for a formula that would generate all the primes and no composites, even if the sequence is destroyed. Nibbling away even further at expectations, we might hope for a formula that would generate only primes, even if many were left out of the sequence. Reducing our expectations even further, we might have to settle for a formula that would generate an infinite number of primes, even if an occasional (perhaps an infinite number) composite were viewed as a necessary evil.

In 1947, the problem was solved at a middling level of desire. More precisely, a mathematician by the name of Mills came up with a formula that would generate primes and only primes (forever) for every substitution of n in a simple formula, though some of the primes might be left out. What was the formula?

The good news is the simplicity of the formula. The bad news, however, is wrapped up with its concomitant humor. The formula is:

$$[A^{3^n}]$$

There is some fixed number A such that each time n is plugged into the above formula, we will be handed a prime number on a silver platter.

The meaning of the "[ ]" is simple. [2.57] = 2; [4.2] = 4; [1.9999] = 1. Thus [x] stands for the greatest integer less than or equal to x.

What a find! Yes, but of course in order to make proper use of the formula, we have to know the value of A. What is it? Answer: Though much effort went into the production of the above formula, there not only is nothing revealed about the nature of A, but furthermore, there is no way of knowing anything about the magnitude of A. A might be some real number (e.g., the square root of 2) between 1 and 5; or it could be a number that is larger than the number of stars in the sky. (For further elaboration, see Brown, 1991; pp. 16–23).

Why is that funny? The incongruity is between what we might expect in the way of an answer to some numerical question in our daily lives versus what we seem to be satisfied with here. If you wanted to know when expected guests would be arriving for dinner, and someone told you that there was a definite time they would arrive, but they could not tell you if it might be within the hour or within the decade, it would of course seem ludicrous.

Assertions that *there exists* something (like some number A having a desired property), even though we may have no idea of its whereabouts, is a kind of statement that is rampant in mathematics, but it is used quite sparingly in most other contexts. We do know of such real-world "existential" statements, but they usually are not terribly helpful. For example, we know (as an empirical inductively arrived at fact) that there is a precise moment at which we will die. That realization may in fact direct the way we live our lives, though it is something we tend for the most part to repress. Without greater specificity, however, it is not the kind of statement that will have much of an impact on how we live in any fine-grained way.

Why does (can) mathematics arrive at such existential statements? The answer is that the special form of proof known as *proof by contradiction* enables us to conclude a great deal that is usually difficult to establish in a deductive way in much of our real-world lives. What makes that sort of proof legitimate is that denial of the existence of something with certain properties leads us to contradict something else with which we might feel comfortable.

Another humorous example in a similar vein is connected with the famous conjecture by the Russian mathematician Goldbach (confided originally to Euler in 1742). Since all primes greater than 2 must be odd, it is easy to see that if you add two primes (excluding 2), you will always arrive at an even number. That is too trivial an observation to get excited about. An interesting What-If-Not application on the nature of the statement however leads to something that has puzzled mathematicians for over two hundred and fifty years. That is, suppose that instead of examining the obvious statement, we look at its converse. The making of such an innocent appearing "tweak" is one way of hypothesizing how Goldbach came up with the following conjecture:

> "Every even number greater than 2 can be expressed as the sum of two primes."

There are some simple instances that seem to support the conjecture, as indicated in Table 5.7.

In some cases, there are two different ways of finding sums of pairs of primes for a given even number. In the above case, for example, we find that in addition to what appears in Table 5.7, we have:

$$10 = 3 + 7$$
$$14 = 3 + 11$$

Goldbach's concern though was not with *uniqueness* of pairs, but rather with the *existence* of a pair for *every odd number*. The funny part now begins. It turns out that no one made any headway with this conjecture for almost two hundred years. Then in 1931, another Russian mathematician, Schnirelman created the first crack in an effort to establish that it might be true. It was in fact a deductive proof. To appreciate what he

Table 5.7 Expressing even numbers in N as the sum of two odd numbers.

4	= 2 + 2
6	= 3 + 3
8	= 5 + 3
10	= 5 + 5
12	= 5 + 7
14	= 7 + 7

proved, recall that Goldbach wanted to establish that at least one *pair* of primes could be produced for any given even number.

What did Schnirelman do? What would be a real funny answer to how many primes you might have to add in order to achieve a given even number, given that Goldbach was hoping to show that at least one *pair* could always be found? Perhaps ten would be funny. Maybe 100, or perhaps 1000 would create a chuckle. Would the need to add 300,000 natural numbers instead of a pair create a belly laugh? That was Schnirelman's contribution to the problem.[15] He showed that given any even number, we can find at most 300,000 primes that must be added in order to achieve it.

So, the ability to come up with something of a deductive nature—one that goes beyond the need to accumulate data—is found to be appealing in mathematical discourse. The incongruity between some sense of what is needed in "practical" situations and what is deemed to be appealing in mathematics is frequently a source of high comedy, and such comedy is often rooted in "existential" mathematical proofs.

Taking Stock and Looking Forward

Our purpose in this chapter has been twofold: to reconstruct the concept of model in connecting mathematics with the real world, and to seek a kind of connection that transcends the notion of model altogether. In both cases we have found it helpful to widen the domain of connection between mathematics and the "real world" by introducing extensions and comparisons from nonscientific fields such as ethics, literature (as in the case of metaphor), and humor. The progression from Figures 5.1 to 5.4 depicts our first efforts at reconstruction, whereas Figure 5.5 sets a different sort of agenda.

With regard to models, we have attempted to explore a truism of applications: that the more we seek and discount irrelevancies from the real world, the easier it is to apply mathematics as a way of understanding and predicting events. It is not so much that the truism is wrong, but rather that its emphasis is limiting from an educational perspective. We have argued that to the extent we wish to see mathematics as a force in the world and not as existing apart from it, the more we ought to seek ways of constructing situations and problems differently. Rather than creating simplified models, we need ways of "complexifying" problems and situations of the world. For some of this complexity, new mathematical models might be created; for others, however, it may very well be that mathematical inquiry is only a partial solution.

Focusing mainly upon the "quick fix" of seeking irrelevancies, limits not only the answers we seek but the questions we ask. The problem of the comatose person, for example, surely involves some appreciation for probability. What it does that the example of Anne, the basketball player neglects to do, however, is that it makes us aware of the fact that knowing the probability of an event may in fact be a very small part of what we need to consider in making decisions.

If we limit the situations and problems we choose to calculating probability alone, we focus on the power of mathematics as a tool but at the expense of asking what it is a tool for. There are in fact some good educational programs that do expand considerably upon the rather limited notion of model that I have explored in this chapter. What is often missing, however, is a conversation that includes not only "accurate" depictions of a situation, but one that also reveals the value judgments needed to make wise real-world decisions.

The second concern—seeking ways of transcending models altogether—is in a way an extension of the thinking about "same and different" that ended chapter 4. The focus there, however, was more closely connected with mathematics per se. In this chapter, we have sought comparisons between mathematics and the real world. Though we have not made it as explicit here as we did in the previous chapter, we have implicitly invoked a mantra that ended our earlier journey. Specifically, we have taken seriously the "with regard to" that we analyzed in the last two sections of chapter 4.

An important theme that influences both concerns is a feminist perspective on morality. Constructs such as "context" and "connectedness" apply not only to morality, however, but to the more general notion of relating mathematics to the real world.

In the final chapter, we further explore some of the thinking that inhibits our inclination to see both problem and real-world applications in a more full-blown context, and we discuss some curriculum experiences that might contribute to that expansion.

Chapter 6

Problems and the Real World: Conclusion and Expansion

> The best stories are those which stir people's minds, hearts, and souls, and by doing so give them new insights into themselves, their problems, and their human condition. The challenge is to develop a human science that more fully serves this aim. The question then is not, "Is storytelling science?" but "Can science learn to tell good stories?"
> —Ian Mitroff and Ralph Kilmann,
> *Methodological Approaches to the Social Sciences*

In the previous chapter we focused upon a number of different ways of conceptualizing the connection between mathematics and "the real world." While still maintaining some of those suggestions, our focus in this chapter will be more on what it means for something to be real—what some of reality's qualities are both in mathematics and in the real world. We end the chapter with a discussion of two educational formats that have the potential to enhance the rejuvenated concept of reality. One of them, the novel, is *au courant* in educational practice and research, though not well represented in mathematics curriculum beyond the elementary grades. The other, the Talmudic format, is a bit more novel than the novel though there are a small number of voices, perhaps whispers, that have called for its incorporation in curriculum.

As an entrée into issues about what is real, we turn first to a discussion of the fundamental belief that lurks a millimeter beneath the surface in a number of arguments for seeking connections. It is the belief that the real world offers an important concrete quality that mathematics lacks. It has an existence that makes it "more real" and thus more accessible both to youngsters who think more concretely than adults and even to mature and sophisticated adults whenever they first encounter a concept.

Such a point of view is consistent with the educational adage that in teaching, it makes sense to proceed from the concrete to the abstract.

While there may be significant pedagogical value in taking that adage seriously, the relationship between abstract and concrete is much more problematic than we usually acknowledge.

Some Tensions Between Abstract and Concrete

In this section we will examine two ways in which the relationship between the abstract and the concrete is problematic. These are meant to be merely illustrative enticements rather than exhaustive analyses. In our first attempt to unearth the problematic nature of the distinction, we will explore a concept that is frequently taught in a concrete way: multiplying positive and negative numbers. In so doing, we will show that there is much more in common between the abstract and the concrete embodiment of an idea than is generally appreciated. In our second attempt, we will show that there is an arena within which there is considerably more incompatibility than is generally appreciated.

Abstract and Concrete: Unacknowledged Commonalty

This example is particularly interesting from an historical point of view. In chapter 1, we mentioned that though there were critical voices of the "new math" during its heyday, for the most part they were resisted for quite a while. One of the most articulate critics was Morris Kline, a prominent research mathematician from New York University. Let us show how he criticized the teaching of a particular concept in the "new math": signed numbers. We will use the deductive teaching of the concept during that era together with Kline's criticism in order to reveal the problematic nature of the relationship between abstract and concrete.[1]

One of the earliest topics that students find to be anti-intuitive involves the extension of the number system from the natural numbers to the integers (the negative and positive numbers). Why it is that the product of two negative integers should be positive is a conclusion that appears on first blush to be more "irrational" than "natural." For ease of exposition, we take the simpler and less perplexing case of the product of a positive and a negative integer to illustrate our point about the problematic nature of the distinction between concrete and abstract.

Below is an explanation found in a ninth-grade algebra text in the early 1960s. This text is part of the series developed by a group of mathematicians and teachers known as the School Mathematics Study Group (SMSG). We have made use of the distributive principle throughout this book and in fact used a form of it to demonstrate in chapter 1 what is

Problems and the Real World

meant by a discovery exercise as part of a deductive framework for the new math. A form of the property asserts:

For all x, y, z, x · (y + x) = x · y + x · z.

Relying heavily upon that principle, the SMSG (1961) explanation for why (2) · (–3) should be –6 is as follows:

(2) · (0) = 0, because of the *multiplication property* for 0
(2) · [3 + (–3)] = 0, by writing 0 = 3 + (–3)
(2) · (3) + (2) · (–3) = 0, if the *distributive principle is* to hold for signed numbers
6 + (2) · (–3) = 0, since (2) · (3) = 6.
Now since (2) · (–3) must be the opposite of (6), it must equal –6.

Morris Kline (1966b), criticizing the SMSG approach on grounds that it presents mathematics as an axiomatic, ahistorical discipline comments, "Extending the distributive law to negative numbers will be of no help at all in understanding negative numbers. . . . The essential point here is that we human beings agree to operate with negative numbers in such a way as to make our formulas more useful." He then proposes that we teach rules for multiplication of signed numbers by relying upon an imagined real-world example. He proposes using the example of temperature rising in a room at the rate of five degrees an hour—assuming that the temperature is 0 degrees at the time of the start of the experiment (0 hours). He depicts the relationship as T = 5 · t, where T is the temperature after the passage of t hours. The issue of negative numbers is introduced by asking for the temperature three hours before the experiment began. We are thus led to consider 5 · (–3) and supposedly would be inclined to conclude that it is –15. His argument is based upon extrapolation from Table 6.1:

Table 6.1 Extrapolating from the product of two positive numbers to find the product of a positive and a negative.

$$
\begin{aligned}
5 \cdot (3) &= 15 \\
5 \cdot (2) &= 10 \\
5 \cdot (1) &= 5 \\
5 \cdot (0) &= 0 \\
5 \cdot (-1) &= ? \\
5 \cdot (-2) &= ? \\
5 \cdot (-3) &= ?
\end{aligned}
$$

To replace the question marks in Table 6.1, note that in each case as we proceed (multiplying by 5 each time) from one positive integer to the next lower one, the product is 5 less. Then if 5 · (−1) is to be 5 less than 5 · (0), it must be −5. Such reasoning suggests that 5 · (−3) = −15.

Now, how does Kline's proposed explanation, using at least a quasi real-world example, compare with that of SMSG? It looks as if they are miles apart. One is based on an axiomatic deductive structure and the other on an act of induction that depends upon a pattern.[2] Do they share common ground? It is perhaps easier to analyze this question if we ask a rather obvious question with regard to the SMSG explanation. Why *must* the distributive principle hold in the extended system? In particular, why must (2) · [(3 + (−3)] = (2) · (3) + (2) · (−3)? The answer is simple: there is no *logical* reason. We are motivated by an *aesthetic* principle which we might characterize as a "wishful thinking" argument. The aesthetic principle asserts: "It would be nice if many of the properties we had before held in our extended system." In fact, it is not that we are merely *using* a property (the distributive property) that has already been established in the set of positive and negative integers. Rather it is that we are *forcing* that property on the new system. It is not that it belongs there in some natural way, but rather we are giving it a home it did not heretofore have.

At first glance it looks as though Kline is presenting an entirely different kind of justification. Is it not "intuitively" obvious that if T = 5 · t relates time to temperature, then the temperature 3 hours ago ((5) · (−3)) must have been −15 degrees? The answer is: There is nothing predestined about the relationship between 5 · (−3) and −15 in this setup. A graph of the linear relationship T = 5 · t will enable us to clarify the point. Let us first plot a few points for the case of the positive integers. Now what should 5 · (−3) be if we do not yet know how to multiply signed numbers? The answer that Kline suggests is illustrated in Figure 6.1.

What does that picture argue implicitly? The graph makes a plea for the continuation of a *linear* relationship as we progress through the negative numbers? Why must the points along T = 5 · t for the case of negatives remain collinear with those for which t *is* positive? The answer once again is that, although this need not logically be so, it would be nice if the properties we ascribe to positive integers were to remain intact as we extend to a new set. Again, the argument is based upon wishful thinking. If we want collinearity preserved, then (5) · (−3) must equal −15, just as we argued in the SMSG case that if we want the distributive principle preserved, then (2) · (−3) must equal −6. There is nothing, however, that requires that we preserve the extension this way. In fact, a pattern which

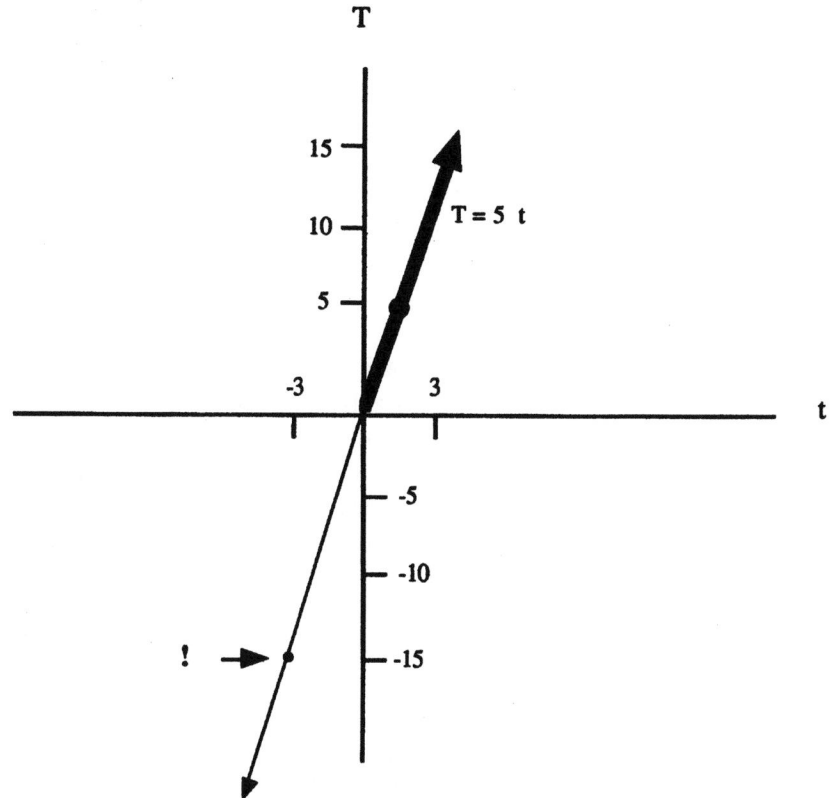

Figure 6.1 Graphical representation of the extension in Table 6.1.

creates a "V" shape, for example, rather than a linear relationship at the origin of Figure 6.1 would be as reasonable as the extension that has been advocated.

What is the point of this elaborate comparison? Although there may be some important distinctions to make between an explanation that is supposedly strictly deductive and one that is supposedly inductive—taking a bow to the "real world"—they share a significant abstraction. The abstraction of wishful thinking and its relationship to proof is a significant one. It may very well be that the subtleties of argument by wishful thinking represent an intellectual hurdle for the student that is itself more overpowering than the distinction between axiomatics and inductive patterns. Furthermore, though all sorts of analogies might be helpful, wishful thinking is an intellectual *tour de force* that cannot itself be made concrete.

We simplify the role of language and thinking considerably when we overlook the fact that even if the real world has an existence in some sense that is different from the mathematical world, the distinction is essentially one of lesser abstraction. We do not claim, as Kline does, that one sort of explanation leads to better understanding than the other. Rather, we have exposed an epistemological issue that would be unwise to ignore as we taut the use of real-world connections on the grounds that such use lowers the bar on levels of abstraction between the two approaches.

Abstract and Concrete: An Incompatibility

The fact that concrete arguments or realizations do not necessarily lower the bar on levels of abstraction inherent in a purely mathematical structure, poses our first conceptual difficulty in using supposedly concrete real-world objects as instances of more abstract systems. Having highlighted a difficulty that derives from what is in common but often hidden between the real world and mathematical abstractions, there is another problem that has to do with the inverse of that one. We now explore what may appear at first to be common between the two phenomena, but is significantly different. It is different in the sense that it renders the two systems incommensurable.

A major difficulty arises when we act as if the real world and mathematical structures are comparable with regard to *existence*. That is, the real world supposedly has an existence that is more easily grasped than a comparable mathematical abstraction. The difficulty is that existence in mathematical terms frequently exhibits a particularly fragile and non-real-world-like counterpart. In an important sense, one cannot be reduced to the other. There is an entire style of proof in mathematics that relies heavily upon claims for existence, for example, that does not share an essential quality with the concept of existence in everyday parlance.

We have already seen examples that are close to satire regarding the concept of existence in our discussion of number theory in chapter 5. There we claimed that a major breakthrough in seeking a formula that would generate only prime numbers resulted in the quite simple, $[A^{3^n}]$. That is, after centuries of seeking and failing in the search for a solution, Mills (1947) arrived at the statement that there does in fact exist a number A so that for every natural number n, $[A^{3^n}]$ would always yield a prime number (where $[\,x\,]$ is the greatest integer less than or equal to x). As pointed out in chapter 5, we disclosed the irony in that solution. While valid in some sense, it comes with a curse. Although we know that A

"exists," we have no idea what the value of A in fact is, nor even its order of magnitude. Such revelations are connected with proof by contradiction. We have also seen this style of proof in chapter 4 when we reviewed Euclid's elegant but slippery proof that there is an infinite number of prime numbers in N. The proof demonstrated that the assumption that there is a largest prime number contradicts other assumptions that cannot easily be discarded.

The claim then that a number in arithmetic or a point in geometry or a function in calculus *exists* does not necessarily mean that we can *produce* such an item. Rather, proof by contradiction demonstrates that if such an element did *not* exist, then some cherished item (statement, situation, concept, number, property) would be in jeopardy.[3]

We see then that the concept of existence in mathematics is not always compatible with its analog in the real world. Not only do mathematical objects lack "touchability," but more importantly, it may be virtually impossible to locate a specific value or even a range within which the purported existence is realized.

Back to What Is Real

While there may be good pedagogical reasons for maintaining the use of concrete examples and arguments in preparation for or in lieu of abstract mathematical formulations, we have shown that the use of real-world models is potentially problematic from two points of view: (1) such use does on occasion place a greater strain on abstract conceptions than is generally acknowledged; (2) there are occasions on which there is much less compatibility between the real-world model and the abstract formulation, and it may be an illusion that we can seek common ground. We are thus led to wonder what might be real about mathematics without attempting to reduce it to what is concrete. If we can do that, then we might be more inclined to expand what might be "real" about mathematical thinking, beyond its connection with concrete models.

Nozick (1990) explores the general concept of what is real in a promising way by identifying how we evaluate reality. He begins by selecting four evaluative aspects: value, meaning, importance, and weight. *Meaning* and *importance* are relational concepts in that they require some comparison among experiences. *Value* and *weight* are inherent concepts in that they are less dependent upon comparison with other experiences. Though something can have *value* in itself, *meaning* requires relating an idea or experience to other ideas or experiences that exist. Eventually the list of qualities of reality and of aspects of evaluating it mushrooms, and he creates elaborate matrices to coordinate the morass.[4]

While a general discussion of Nozick's work is beyond the scope of this book, we focus on one quite interesting question that may appear less obvious than others. He asks what it is that makes some objects or experiences "more real" than others. Despite the fact that they are fictitious, he chooses literary figures as an example of where we can find reality. Nozick comments,

> Some literary characters are more real than others. Think of Hamlet, Sherlock Holmes, Lear, Antigone, Don Quixote, Raskolnikov. Even though none of them exist, they seem more real even than some people we know who do exist. It is not that these literary characters are real because they are "true to life," people we could meet believably. The reality of these characters consists in their vividness, their sharpness of detail, the integrated way in which they function toward or are tortured over a goal. Even when their own focus is not completely clear, they are intent on focusing or are presented (as Flaubert presents Madame Bovary) in clear focus. These characters are "realer than life," more sharply etched, with few extraneous details that do not fit. In the characteristics they exhibit they are more concentrated centers of psychological organization. Such literary characters become by words, paradigms, models, epitomes. They are intensely concentrated portions of reality. (p. 130)

Speaking of works of art, he comments,

> Works of art, paintings or music or poems, often seem intensely real; their sharply etched features make them stand out against the usual background of blurry and vague objects. In a mode of organization more tight and coherent, or at least having a more evident mode of organization and a more interesting one, they constitute more integrated wholes. The beauty of works of art or of natural scenes, the dynamic balance of the array, makes it more vivid, more real than the usual jumble we encounter. Perhaps this is because beautiful things seem right as they are; they show a perfection of their own. Or perhaps it is because, on their own merits, they hold and repay our attention more enduringly. In any case, they are perceived as in sharper balance and focus; they are more vividly perceived. Features other than beauty, such as intensity, power, and depth, also lead to vividness of perception. Artists are trying, I think, to create objects that, in one way or another, are more real. (p. 130)

Moving from literary characters and works of art, he speaks about some famous people who are more real than others: "Socrates, Buddha, Moses, Gandhi, Jesus—these figures capture our imagination and attention by their greater reality. They are more vivid, concentrated, focused, delineated, integrated, inwardly beautiful. Compared to us, they are more real" (p. 131). In all these categories, what accounts for greater reality, even when their existence in traditional terms is questionable, is that they capture our imagination; they are more interestingly organized than more humdrum counterparts; they are more vividly perceived; they are perceived in sharper focus and balance; they are more aesthetically appealing than objects that are "less real." It is in an effort to rejuvenate a loss of vitality and connectedness that typifies our less real lives, that Maxine Greene (1978) amplifies the concept of "wide-awakeness" based in part upon the work of Henry David Thoreau and Alfred Schutz. She comments, "We are all familiar with the number of individuals who live their lives immersed . . . in daily life, in the mechanical round of habitual activities. We are all aware how few people ask . . . what they have done with their . . . lives, whether or not they have used their freedom or simply acceded to the imposition of patterned behavior . . . ."[5] We turn now to an application of this enhanced concept of reality to the field of mathematics education.

Reality and Mathematics/Education

If we can speak of what is "real" in a more vibrant sense than what "exists" or what we can "touch" and "see," then we not only legitimize more interesting connections between mathematics and the real world (much of what we did in chapter 5), but we also suppress the need to seek real-world connections as a salve against an otherwise "unreal" world of mathematics.

Why is it, for example, that the Greeks were so captivated by irrational numbers? Why might students of the twenty-first century (or perhaps the eleventh, if they are victims of a Y2K problem) be similarly enthralled? As with some literary characters or with great works of art or with people who are indeed more vivid than others, these numbers raise our sites considerably on what a number might be. Set against an expectation that it would be hopeless to be able to find any such "creatures" on an interval on the small stretch of the number line that is one unit long, we find out not only that there is *one* such number—$1/2 \cdot (\sqrt{2})$—and not only that there is an infinite number of them, but in fact there are "more" of them than of the rational numbers in that interval, or any such interval, no matter how small!

Implied in the above discussion is the "reality" of an infinite set. None of us has ever seen or touched an infinite set, but we may have dreamed in our youth of what it would be like to reach the end of the universe and to wonder if we would fall off or if we would continue forever. As with irrational numbers, the concept of infinity makes the idea of number itself a much grander enterprise than we could have imagined. To try to find ways of talking about operations on such sets puts in bas-relief what we previously understood about more mundane systems.

Another category that has the potential to increase both meaning and importance of objects in the mathematical world is that of paradox. We have already discussed how Russell's paradox of sets (in the discussion of self-referentiality in chapter 5) was not only a "cute" addition to the logic of sets but required a total revamping of what we previously took to be a nonproblematic concept—the notion of set itself. But there are many more accessible paradoxes that are more easily resolved, that still have the potential to heighten our sensitivity to what we had previously seen as lacking in surprise.

Problem (1) of chapter 4 in the section entitled "Unearthing Implicit Ingredients of Comparison" is such an example—a "relatively" simple problem but, as we mentioned, one that intrigued Einstein. The question "Why can't we travel back down a path at a fixed speed so that the total average is twice the speed of the way up?" places the concept of finding the average speed of two average speeds in a new light.

But there is more that makes something "more real" than finding out about the extraordinary. As we discussed with regard to wonder and surprise, the best sign that our ability to wonder has ceased is that we no longer seek to uncover it in our "everyday world." As soon as we see objects and relationships *as if* they could be or could become something else, we have bestowed upon them greater importance, significance, worthwhileness than we had seen before. We have, in short, given them a greater sense of reality.

In preparation for a discussion—in the final subsection of the book—of two educational perspectives that have the potential to integrate much of what we have said about both problems and connections with the real world, we turn in the following two subsections to a brief summary and consolidation, respectively, of the two major foci of the book. As a summary of sorts, these sections should provide a quiet respite and an opportunity for fortification in preparation for the final hurrah.

A Retrospective on Problems and Connections

While it is possible to view this book as a criticism of *The Standards* and associated documents, I have been less concerned with an analysis of the proposals of these documents than with using them as an entrée for the exploration of ideas with regard to the two dominant themes—problem solving and applications—that have been suggested by them. As must be clear by now, much of my thinking about these matters predated *The Standards* by several decades, although some of it has taken a different cast as I profited from insights and shortcomings of these documents. I would be less than candid, however, to claim that I needed *The Standards* for all that I wanted to say.

It has nevertheless been helpful to use these documents to both clarify and "muddify" much of my thinking. There are points at which what I am suggesting is a sympathetic criticism/extension of that literature. There are others for which I appear to be operating in a different paradigm.

Some of the discussion of problem posing appears to be in a sympathetic mode. The strategies associated with What-If-Not thinking can be viewed in that spirit. I have, however, attempted to excavate the concept by locating the various ways in which it is an essential part of thinking and understanding, and not merely a motivational fad or add-on to the process of thinking.

Other orientations, however, that I have focused upon may be less compatible. Seeing problems as having the potential to reveal important aspects of self, of others and of the *nature* of thinking/feeling, makes use of mathematics in a different spirit than is generally called for in *The Standards*. Here problems and problem posing/solving are used not only to motivate and to clarify mathematical bodies of knowledge, but also to see that body in relation to a host of other ideas. Some of the ways of thinking in chapters 4 and 5 may not only be different from, but are antithetical to the concerns of these documents.

There are some interesting gray areas in the category of compatibility. The grayness has to do in large measure with the intended educational purposes served by the analyses of problem, problem posing, and problem solving. One such gray area most likely surrounds the various sketches I have suggested throughout. Without rehearsing in great detail the analyses in the previous chapters, a review of Figures 4.1 to 4.5 will serve as a valuable reminder of the reconstructed attitude towards problems and problem solving in mathematics education that I have attempted to develop.

Figure 4.1 reminds us what the standard fare of problem solving has been for a long time and serves to inspire us to think otherwise. Figures 4.2 through 4.5 offer us the beginnings of some "otherwises." The role of context in Figure 4.2 began our investigation of why it is we might inquire about anything. Figure 4.3 on *situation* as a generator and receiver of problems contributes significantly to a humanistic point of view.

In terms of a more standard problem solving curriculum, even when problems are presented for the purpose of seeking solutions, the invitation to "neutralize" them is one that will help clarify what it is that the problem is demanding. To unearth the implicit demand in a problem is to unmask what makes it a problem in the first place. Disagreements about different perceived or implied problems and accompanying demands that reside in situations are part of a more radical conception of what it means to be humanistic. Dialogue that uncovers why we perceive these differences, enables us to better understand how and why each of us sees the world differently.

In chapter 3, we analyzed the velvet hammer that had been introduced in chapter 1 through a variation of the game of jeopardy. The educational fallacy essentially confuses the *educational uses* to which problems might be put with the *definition* of problem itself. The identification of the fallacy alone does not necessarily set an educational agenda. If the educational agenda set by problems is not limited to that of seeking their solution, then how might they be used? I see the answer to that question as the start of a significant debate among educators and between educators and their students. Though I have attempted to provide an answer to this question from the perspective of humanistic education (that has led to much of my analysis in chapters 2, 4, and 5), there are alternative perspectives, and I would be happy to see this question be used in the service of uncovering and debating other platforms as well.

The schema of Figure 4.5 is an excellent, if not perhaps a bit confusing, example of how we might be affected by the realization that problems do not necessarily have to be used for solving alone. That is, one interpretation of the constructivist paradigm suggests that to "give" both a problem and its solution essentially robs students of thinking on their own. While such an approach might be justified on occasion, mounting a program with that orientation would leave little to engage the student (so the argument might go).

We have shown in chapter 4 that there is much to think about with regard to mathematics, self, and culture, even with what looks like a traditional teaching paradigm. A pressing question is: What analyses can be derived from the scheme of Figure 4.5 as a starting point? Again, nothing *automatically* flows from the realization that solving of a problem is not necessarily the educational purpose for which that problem might be used. A clearer rendition of what we have suggested in Figure 4.5 was found in Figure 4.6.

Figure 4.5 was included in chapter 4 in order to remind us that there can be a great deal of valuable inquiry that follows upon a solution to a problem, even if it is "given or imposed" by someone other than the student. The inquiry can be closely connected with mathematical exploration, but it can also move in other directions, ones that are metamathematical or ones that serve to connect mathematical thinking with other domains.

In chapter 5, we began the exploration of the second major theme of the book: connecting mathematics with the real world. There we showed that connections can be in the form of models, and that is what most curriculum views of connecting mathematics to the real world portray. The remainder of that chapter works out other types of connections, ones that push the bounds of the concept of humanistic beyond what is found in *The Standards*. We not only offered a reconstructed notion of model, but attempted to transform the *kind of* connection being sought. In investigating how mathematics compares with humor and morality, for example, and how it makes use of a device (metaphor) normally associated with poetry, and other forms of literature, we are captivated by how it is that *domains* relate to each other rather than by how one can be used in the service of another.

By analyzing this kind of connection, we make use of a construct that we introduced earlier with regard to problems, i.e., the notion of similarity and difference. What is interesting is not *that* similarities emerge among apparent differences, or vice versa, but rather what we have called the "with regard to" modifier of similarity and difference. We examined various ways how this modifier might be used with problems. To be ever so slightly self-referential, the reader may find it enlightening to seek out how these modifiers (and others) play themselves out "with regard to" connections themselves.

We took a final stab at the concept of connections in this chapter as we moved from "real-world connections" to exploring what it is that makes something real. We took this detour mostly to challenge the prevalent pedagogical truism that suggests that we need real-world connections as a salve against the perception that mathematics is a body of knowledge and experience that is, relatively speaking, lacking in reality. We choose the real world (including, but not limited to manipulatives) supposedly as a buffer in preparation for handling the collection of mathematical abstractions.

This is not the only argument for drawing connections with the real world, but to the extent that it is taken as an important pedagogical motivator, we have found it worthwhile to temper the truism. In particular, we have shown that what it is we wish to draw from the real world is frequently not merely a concrete embodiment of an abstract mathematical idea but rather a significant abstraction in its own right, an abstraction that may be hidden in focusing too closely on what is considered to be concrete. The example of using temperature as a real-world motivator for developing a rationale for the multiplication of signed numbers is a case in point. There is nothing particularly concrete about an alleged proof that depends upon "wishful thinking" for its justification. Yet, once the dust is cleared, it turns out that it is this principle which is operating in *both* the axiomatic and real world interpretations of multiplication of these numbers.

Conversely, we have focused on ways in which it may not be possible at all to find real-world counterparts for mathematical abstractions since they may involve levels of abstraction (such as the concept of *existence*) that are incompatible in the two domains. None of this proves that seeking connections between mathematics and the real world is a bad pedagogical idea. It does, however, open up many more avenues to explore than we generally acknowledge. Wondering about what there is that is really "real" about mathematics and our experiencing of it leads us not necessarily to give up efforts to make connections with other domains, but it leads us to think differently about that body of knowledge itself. On the other hand, seeing what is abstract in supposedly concrete experiences leads us to appreciate how much closer a bond there may be between mathematics and the real world especially when the latter is used as an inoculation of sorts against the former.

Three Resituated Perspectives

There are many points of view that accompany the reconstructed view of problem and connections of mathematics to the real world. In this section, we highlight three that have been mentioned in passing, but they bear upon both major themes of the book. The three are: pseudo-history, error-making, and the phrase "thinking like a mathematician." Of the three, the third is the most all-encompassing. It is particularly important because, unlike the first two concerns, criticism of it has slipped between the cracks of both the "new math" and *The Standards* movement.

On "Pseudo-History"

Much of what we have discussed in the retrospective section above offers the potential for transformation, seeing things not as "given" forever, but as if they might be otherwise. As we argued throughout this book, one valuable member of the tool kit for doing so is that of What-If-Notting.

We have alluded to another (perhaps a variation of What-If-Notting) in several places as "pseudo-history." Actually it appeared implicitly in subtle ways with little fanfare in a number of places. One such place was in our discussion of Goldbach's conjecture. That is, we frequently operate as if historical accuracy (even if watered down) is necessary if we are interested in incorporating the evolution of ideas in classroom settings. In fact, for many purposes, we can find value in imagining what *might have* been responsible for such evolution.

In the case of Goldbach's conjecture, I imagined how he might have come up with the conjecture in the first place. Thus I thought about other conjectures that appeared to be connected with his assertion that any even number greater than 2 can be represented as the sum of two primes. There were many possible ways of imagining forerunners, but one that struck me as particularly revealing was the *converse* of that conjecture. That the sum of any two primes (excluding 2 for the moment, though it might be incorporated in the dialogue) is even seemed so simple that almost nothing was needed to prove it. I imagined that Goldbach might have started with the simplistic converse and wondered what kinds of challenges could transform such a trivial observation into something worth thinking about. Ergo: Goldbach's conjecture. If nothing else, a What-If-Not mentality has taught me to be cautious of dismissing apparently trivial observations. All trivial observations are a heartbeat away from ideas that are significant.

Pseudo-history has become a commonplace in many fields that appear to be lacking in documentation of the past. Some feminists for example, have found it helpful in uncovering a sense of womanhood, to create a tapestry of their own history based upon threads that are personal rather than derived from scholarly sources. In that case, many women find it helpful to look inward as a community in order try to understand how women experience the world and how they learned to define themselves. This kind of activity can profitably engage students in all fields of inquiry.

Borrowing from the feminist model, the classroom of the present can be a laboratory for history. That is, students can learn to keep a record of how their thinking/feeling evolves over time. What is real then is not what history as an official document stipulates, but rather how we create it from within. It is a pretend history of sorts, but nothing could feel more real than to act as if we had created the history based upon our own lives.

Making use of pseudo-history in relation to history imposes an interesting phenomenon on the template of Table 1.1—the ambiguity of humanistic mathematics education. Making use of history in and of itself (in an informative sense) would seem to fit into box A (or perhaps B). Pseudo-history, however, if used as a pedagogical ploy would seem to be an act of construction—and would belong more comfortably in box D.

In such an environment, students could not only do explorations and reflect on them, but this activity could be part of the historical record of the course itself. A significant part of the curriculum would not only deal with problem solving, heuristics of problem solving and looking back at what they "discovered," but would help students recall the debates and differences of opinion that were expressed over a protracted period of time as the students themselves struggled to make sense of new ideas.

Part of what is needed in order to incorporate the evolution of ideas as an element of curriculum per se is an inclination to appreciate the various sized chunks and kinds of chunks of inquiry. To view a problem to be solved as a chunk is all well and good. But there are chunks of other sizes and kinds. In the category of *size*, we might consider not single problems, but collections of problems, entire units, and even mathematics as a field. The sorts of questions we pursue might be quite different depending upon the size of the chunk. In addition to size of chunk, the issue of *kind* raises the question of what else there is of value to explore other than problems per se. We return full circle to problem posing, situations in relation to problems, the relation of it all to "self," and other such themes we have explored throughout.

All of this suggests a quite different notion of pedagogy than one in which students work on a problem (even if done in groups) for an hour or so, come up with different solutions, share and evaluate them and then move on. It requires, for example, a much greater tolerance for errors and error making and a more tempered view of correcting them.

On Error Making

In seeking to imagine the evolution of ideas, it is necessary to be cautious about the handling of errors, a theme we alluded to in connection with non-Euclidean geometry. For a long time, it was popular for educators and researchers to take a quite narrow view of what it means to "learn from your mistakes." It essentially meant that mistakes were to be identified and then eliminated. Important concomitants of this orientation lay in the areas of diagnosis of errors. Specifically, teachers learned necessary skills in order to identify what errors their students were making and to figure out what misconceptions may have been responsible for such errors. At best, teachers would then modify their teaching regimen and attempt to eradicate such error making.

A more interesting paradigm is emerging, and one that is a healthy supplement to what we have described above. That is error making can be viewed in a generative sense. It can serve the purpose of raising new territory to explore. Questions like the following have the power to transform reality from what is given and settled to what might be:

i   What new questions do my errors encourage me to ask?
ii Which errors might be worth exploring? What might I be able to find out if I do not immediately discard what might be an error?
iii Is there any context (perhaps a different question from the one I am exploring; perhaps a different system from the one I am investigating) in which my errors might be at home?
iv For what experiences in my life does the concept of error make sense? For what experiences does it not seem to apply at all?

In order to view error making in a generative sense, what is needed is not so much a new curriculum as it is a change of attitude or point of view. In this regard, a critical look at the following account by Fielker (1990) is a helpful beginning. He describes how error-making figures into an innovative teacher education model derived from what Caleb Gattegno, called "observation lessons." In making use of a special form of model teaching, he points out a number of principles that govern his performance with youngsters. Among the principles are:

i   Never tell the children anything they can deduce for themselves,
ii Do not correct mistakes. (Fielker, 1990, p. 19)

In pointing out that as a teacher-educator, using this one-shot model, he does not have the leisure to postpone corrections that teachers might have in a more standard teaching set-up, he remarks, "In the classroom one can decide to leave it to the next lesson, or in a secondary school, set an appropriate homework! But an observation lesson has no follow-up, so that very often a decision has to be made about letting a strange group of children leave without something being corrected. . . . Luckily the situation rarely rises, because usually the children discover errors before the class finishes" (p. 21). Why are these students with whom he is unfamiliar so adept at catching their own errors before the lesson ends? He comments, "Often I feel this is due to subtleties of teaching that are difficult to analyze: the more pointed question, the gesture that changes the focus of attention, a steering onto a different track that will lead back to the error, a fresh example that will lead to a contradiction, or even a blatant query about whether something is correct" (p. 21).

Now what Fielker describes above may be excellent examples of tacit knowledge that "good" teachers have learned to employ even without explicit instruction. But such tacit knowledge says volumes about beliefs and attitudes with regard to the content of curriculum and about theories of "coming to know" something new.

Much of education may legitimately be about passing along a well established body of knowledge together with concomitant theories of encouraging students to "discover" that knowledge on their own. Nevertheless, it is important to appreciate that even with this goal in mind, we need to distinguish between "coming to know" that body of knowledge and the knowledge that is eventually acquired. There is considerable value in intentionally designing educational experiences so as to suppress what is true and even to encourage false beliefs.[6]

We may on many occasions want to set students back on track. However, unless we allow for and even encourage error making, we not only interfere with the opportunity for students to explore uncharted territory that could even clarify our own predetermined goals, but we prevent them from taking a pseudo-historical perspective on their own growth. Such a point of view discourages them from appreciating that the history of ideas is a labored history that demands debate, controversy, and confusion, each of which influences not only what but how ideas will be explored.

More importantly, the focus on errors, whether corrected or not, has some built-in assumptions about both the manner and the content of instruction. Consider the following (from Brown, 1990):

> Errors and mistakes become an important category primarily when the concepts of correct, right, proper, and true dominate our curriculum concerns. . . . The dominance of these categories in mathematics classes suggests that important parts of one's emotional and intellectual education are being neglected. What are the kinds of issues, problems, situations, questions that might legitimately be explored in the context of a mathematics classroom for which the concept of "being mistaken" or "committing an error" might not apply at all? (p. 37)

In answering the last question, we return full circle to the major themes of this book. Given a problem, what can be done other than solve it? What kinds of metaphors do students use to construct or retrieve ideas? How does our experiencing of mathematics relate to the ways in which we experience other fields? Given several different solutions to a problem, which do we find aesthetically more appealing? Furthermore, issues of emotionality are not only activated by "right/wrong" radar. The questions and issues abound. Borasi (1986, 1987) has explored some of them in the form of teaching experiments. At this point, though, we leave it to the reader to apply themes we have developed and also to create new ones for the purpose of clarifying how error making may be a generative and essential ingredient of humanistic education.

"Thinking Like a Mathematician": A Critique
Much of what we are projecting in a radical humanistic mode is visionary and requires not only further criticism and analysis, but imagination and courage before constructing an educational program. It is easier to advocate than to put in practice a view of mathematics education that supports

1) the classroom as open dialogue;
2) the holding of resolution in abeyance;
3) the potential of errors not only to diagnose misperceptions but to generate new territory;
4) having students record the evolution of their thinking over a protracted period of time;
5) integration of mathematical experiences with other fields in a more global and pervasive manner than is suggested by "applications";
6) using the discipline to reflect upon an understanding of self and society.

Though researchers over the past few years have engaged in enlightening ethnographic-style research in seeking how teachers and students acquire meaning of fundamental educational and mathematical concepts, they have tended to cast a narrower net than is suggested by the array of concerns we have associated with a humanistic orientation.[7]

In addition to rethinking what forms of teacher education might be needed to support such a program, there is an interesting issue regarding the conception, uncovering, and support of mathematical talent, especially genius. Although we are not focused on educating such a population, the conception of such talent tends to have a "trickle down" effect because it represents an implicit ideal. In chapter 1 we raised the issue of what constitutes talent in mathematical thought, and we elaborate upon it here.

In Krutetskii's (1976) classic work on mathematical talent, he makes use, among other things, of "think aloud" techniques. He asks students to share their thinking in the process of working on problems. He points out that seeking clarity, simplicity, and elegance in solving problems distinguishes the youngsters who excel from others. Furthermore, aside from problem solving per se, information is both acquired and retained in an economical manner so that the pupil is not overloaded with surplus information.

Such talent is revealed in a manner that in many ways is antithetical to the schemes we have presented here, schemes that require thinking deeply about complicated and personal matters. Furthermore, they focus less on coming up with a clever problem solving dénouement and more on appreciating the source and intention of questions, as well as their possible unanswerability. What we have proposed in the way of drawing connections with the real world shares more with the worldview associated with a philosopher and a novelist than with the one of mathematician as portrayed by Krutetskii.

Much of what we have pursued, however, enables students to have a better appreciation and understanding of even standard mathematical topics. In addition, though, it paints a picture of an educated person that far outstrips the *weltanschauung* that has been associated for decades with the darling of curriculum innovation in mathematics education: *thinking like a mathematician*.[8] While we may buy something worthwhile by a radical humanistic program in mathematics education, a better educated person, it is worth pondering what might be jeopardized if a scheme of the sort proposed in this book were to supplement a curriculum that was more centrally focused on *thinking like a mathematician*.

Although we may, for a variety of reasons, wish to have ways of identifying a narrow conception of mathematical genius, we need to be educating our population at large with a different sort of agenda in mind, an agenda which encourages all of us to be wiser, more aware, more critical, and more creative not only about individual disciplines but about the nuances that connect those disciplines to greater understanding of self, others, and society.

"Novel" Endings

There are many ways to think about how we might create an environment that could encourage a humanistic encounter with mathematical thinking. Surely, it would be a mistake to rely upon text as the only provider of such services. There are some innovative computer programs that encourage forms of problem posing such as Turtle geometry and Geometric Sketchpad. In addition, we have suggested throughout how the social context of the classroom might mimic in important ways some of the emotional and intellectual components of the evolution of a subject.

Yet there is much that can be done with texts that have the potential to construct a variety of humanistic environments (box D, but perhaps box C as well in Table 1.1). We will suggest two examples: the novel and a secular Talmudic format. Though there is a precedent for the former, there is value in thinking through why that genre might work for our purposes and how it might be modified so as to be more consistent with some of the tenets we have been proposing. I am presently engaged in creating materials with a Talmudic format. To my knowledge, there is not yet any other material of that sort that has been developed in mathematics education, though it is a "near relative" of hypertext and other computer based styles of inquiry. I select these two formats, the novel and the Talmud, since one is an example of an emerging genre while the other is promising but still at the drawing board stage.

Let us begin by wondering about the implicit messages of texts that make them such poor candidates for a humanistic orientation. Some of these messages are the result of poor technique. Others seem to come close to being built into the genre itself. At any rate, the following slightly caricatured rendition of text might suggest valuable alternatives.

Sequentiality and careful logical development are a text's hallmarks. Connected with linearity, however, is a general lackluster "story line." It is rare, for example, that readers are encouraged to figure out the relative importance of one idea in relation to others. Furthermore, it is usually assumed that the purpose of the text is to familiarize students with information or skills about which they supposedly know little or nothing. It is not only that they lack language to describe what will unfold, but rather that they supposedly possess no previous independent experience to enable them to appreciate the topic. There is also the implication that it is the text's responsibility to be unconfusing and to provide clear explanations. Behind this assumption is that the ideas presented not only are uncontroversial, but that they were born full blown with no labor pains to boot. In short, everything, including problems to be worked on, is "given" with little context for its evolution and little awareness that much is contestable. Most textbooks, especially in mathematics at the post-elementary school level, convey little in the way of conflict, unsolvability, drama and emotionality. How can the situation be rectified?

It would be a mistake to isolate the influence of texts from the social/ political milieu of schools and classrooms. Obviously much is in need of repair with regard to the authoritarian, undemocratic nature of the school in order that a revision of text have a significant influence. There is much that is taking place in the reorganization of schools that offers hope that innovative texts and social climate will not be at odds with each other. As far back as 1971, Sarason identified the failure of the "new math" with the mismatch between a less authoritarian curriculum and a highly authoritarian school setting. To some extent we are in the throes of taking seriously the following insight about a democratic school environment and the teacher's judgment about educational matters. This 1931 insight of Dewey's has a quite familiar twenty-first century ring: "Until the public school is organized in such a way that every teacher has some regular and representative way in which he or she can register judgment about matters of educational importance, with the assurance that this judgment will somehow affect the school system, the assertion that the present system is not, from the internal standpoint, democratic seems to be justified" (Dewey, 1931, p. 199). As an aside, notice Dewey's reference to "he or she" in the above 1931 quote! There is of course more to say about what else might be needed other than innovative text in order to transform educational practice, but let us turn now to the issue of text as novel, and leave some of this important conversation as a topic for further reflection.[9]

First Transformation: The Novel
At about the time that the "new math" movement was peaking in the United States, there was a fascinating study conducted by a pair of anthropologists who went to Liberia to figure out why the introduction of Western society's "new math" into that ancient culture had failed dismally. This venture predated our more recent multicultural agenda—one that would have striven to unearth the mathematical insights that the Kpelle tribe of Liberia had to offer Western society. The question that John Gay and Michael Cole (1967) asked in *The New Mathematics and an Old Culture: A Study of Learning Among the Kpelle of Liberia* was one that would be considered nowadays to be unabashedly ethnocentric. They wanted to know why our Western program, which focused on a careful, logical development of mathematics, was so poorly received by the Kpelle tribe.

The results of the study are too rich to capture in this brief section, but one universally applicable insight that emerged speaks about the role of folktales in this so-called primitive society. The values of this culture are sustained by telling tales about all aspects of the environment. Furthermore, these folktales are passed on with less of a concern for veracity and more for improvisation and wit. As Gay and Cole point out, kudos go to those who leave others speechless by introducing clever and unexpected twists.

How could a program that focused primarily on a logical perspective ever have succeeded in this culture? As in all preliterate societies, the telling of stories was the backbone of the culture. It is an insight, however, that we have begun to rediscover and adopt in modern society as well. Though the dominant view of education is that it is an enterprise in which preordained knowledge is passed on, a much more appealing image has been emerging, i.e., telling and exchanging of stories that inspire the kind of improvisation that is deeply embedded in the Kpelle culture. There is something universal about the Kpelle reaction as is captured by Mitroff and Kilmann's quote on scientific inquiry at the start of this chapter.

The following beginning of a story suggests how it is that a mathematical story might inspire a way of thinking that is at odds with the positivistic worldview of texts we depicted at the beginning of this section.

> Amy was doodling on her pad next to the telephone as she was talking with Joanne about how the world began.
>
> "Don't you believe really that God created the world and everything in it?" Joanne asked.
>
> "I don't know really what I believe." said Amy. "My father has always told me that things have causes, and we have to find those causes if we really want to understand them. But then he tells me that it's important to go to church, and I can't really get him to talk about why it's so important. Do you think he can really believe in God?" said Amy.
>
> "What do you mean?" asked Joanne.
>
> "Like, do you think he believes that God was always there? Like, if God was always there, who created God, and what was the first thing he did? . . . " Amy began sucking on her long blond hair. She noticed that while talking she had been drawing something resembling the picture below:
>
>
>
> Why had she drawn that? At first it reminded her of a part of a barbed wire fence. She was pondering the question of whether or not everything has a beginning, when it occurred to her that the picture looked like a number line that her math teacher had drawn in class the day before. She found herself putting numbers beneath the line:
>
>
>
> She thought of herself walking along that line, and realized that if 0 was her starting point, then 1 would be the first number she would hit: 2 would be the second one, 3 the third one, and so forth.

Then she thought of herself as a bee and the numbers on the line were flowers from which she gathered nectar to make her honey. She looked forward to arriving at each flower with great joy. She would tell the second flower what the nectar was like from the first flower: she would tell the third flower about some fellow bees she met at the second flower. She no longer heard what Joanne was saying at the other end, but she said "Um hum" every so often just to keep the conversation going.

As she thought of the number line, she remembered that her math teacher had recently been doing something with fractions. She decided to put some of these fractions on her number line:

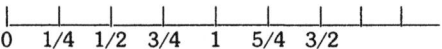

She became a bee again and asked herself what would be the first number she would touch as she moved along the line from 0. That seemed easy, 1/4. But then her mind jumped to another thought: many numbers were missing from her number line. She placed 1/8 in between O and 1/4, and decided to include 1/8 spacing all along the line. But that wasn't enough . . . . She found herself asking what was the first fraction after 0 that she'd hit if she were a tiny organism crawling along the number line. She was very upset by the question. She could not answer it because every time she found a small fraction she was able to find a smaller one.

Then Amy became frightened—in a way that she remembered becoming frightened as a little child—lying in bed at night, and wondering what would happen if she were floating around in the universe and she suddenly came to the end. Maybe there was no end. She didn't know which of the two possibilities was more frightening. Was this like the question about the beginning of the world? She wondered. Was the question a good question? She wondered about that too. What was a good question? . . . All of a sudden she heard a click at the other end. Oh, no! Now she wanted to talk to Joanne, her closest friend, more than ever. Would Joanne be upset with her? She hoped not, but she realized it hadn't been nice of her to forget about Joanne completely, however interesting her thoughts. . . . She should call Joanne back immediately, and explain. After all, Joanne was a friend and she would understand. (Borasi & Brown, 1985, p. 22)

In stories the concept of dialogue is not something we experience *in addition to* reading an otherwise uncontroversial text, but it is built into the story itself. Imagine mathematical novels written in which youngsters work through the different conceptions of infinity they hold and the joy and fear associated with flying through space with no beginning and no end. Imagine, as in the excerpt above, the connections that could be made between the mathematical experience and views about the beginning and end of the universe. Though technical matters about fractions are not irrelevant, the debate in this story line is less on such matters per se and more a search for the first fraction larger than zero. As a matter of fact, issues about "the first" fraction after zero provides a first-rate sort of motivation for learning about such technical matters. Furthermore, it would not be difficult to introduce dialogue in which the search for smaller and smaller "first fractions" is included in the context of a discussion for selecting among a number of different strategies, some of which may be more elegant than others, some of which may be easier to perform, some of which may be grossly incorrect, and yet others that may be wrong but not too far from the mark, and so forth.

Notice here we have made use of a mathematical story that connects with the real world in a way that transcends the use of models in the narrow sense depicted in Figure 5.2. That is, we are seeking a common theme in mathematics and theology. In both cases, we are flirting with some form of the question: What is the first? There is much possibility for debate here. We can explore questions like: If the points along the number line have some thickness to them, then does that affect our search for a first? Are the two situations analogous? The mathematical question assumes a starting point, 0. The one about the universe appears not to. How does that difference affect how we view the two questions in relation to each other?

I have shown this vignette to people of all ages and of different levels of mathematical sophistication. Some of the most imaginative mathematical and "real-world" thinking I have encountered came from people who wanted to continue this story on their own terms. Though I have not done systematic research on what accounts for its success, I imagine that in addition to the mystery and tension both within and between the theological and mathematical realms, they are intrigued by the dynamic and "unreal" bee-a-morphic behavior of Amy.

Many readers picked up on the fact that Amy let her imagination run wild and saw herself communing with flowers along the path. They too wondered what they might say to each flower about their discoveries, feelings, confusions. Some of them spoke about their frustration at not being able to find a "next" flower upon which their "beeing" might alight. They imagined a conversation between themselves and the flowers along the path, wondering whom they might grace next and whom they would disappoint. Some were intrigued by the way in which the interpersonal relationship between Amy and Joanne was woven into the story. In particular, the theme of friendship—and of elements that strains that friendship—inspired several of them to try to understand what caused the breech and why. They also began to wonder when it was that they tuned out on conversations with their own friends and what they saw as the consequence of such behavior.

Friendship—more intricately played out—is an underlying theme in the mathematical novella, *Posing Mathematically*, that I mentioned in the preface.[10] As I disclosed there, during my last sabbatical leave I had been involved in writing it, but was reluctant to share my thinking about it with my philosophy of education colleagues at the time. When asked to give a presentation at a colloquium, I ended up speaking about another topic: the one that evolved eventually into the central themes of this book.

Having completed the novella, I am now prepared to comment briefly on it. The suppressed novella now merges as an element in the very scheme that I substituted for it in an attempt to avoid discussing the novella. Enough playing around the edges of self-referentiality. Full speed ahead with the description that was avoided at the PERC colloquium!

The novella attempts to achieve for teachers and teacher educators similar goals to the ones we described in the above vignette of Amy and Joanne. The story is about the professional and personal lives of two experienced teachers, Martha and Sy, who decide to engage in team teaching. Along the way, they come up with some interesting ideas about problem posing. As they begin to feel more comfortable with their ideas and their personal relationship, they wonder how they might make their ideas accessible to their colleagues. The novella incorporates educational strategies that are designed to invite the reader to criticize and elaborate on the ideas they are developing. A number of activities are created by the characters of the novella and are presented to an "audience" with whom the reader might identify. Such strategies encourage an active type of communication between the protagonists and the reader. The novel format attempts to minimize the sugar coating of a didactic set of messages of the sort that appears in standard text formats.

Interesting tension is built into the story by the presence of others (especially Eloise, the department chair), who find much of what these teachers are discovering to be impractical and indefensible from mathematical as well as pedagogical points of view. Some of her criticism, though, is motivated by a feeling of jealousy over Martha and Sy's relationship and their popularity with their students. The protagonists take that awareness into consideration, but eventually they choose not to dismiss the chair's criticism out of hand. After considering a number of options, Sy and Martha decide to apply for a grant from several agencies to enable them to have released time in order to write a novel as their vehicle for passing along what they have been agonizing over and discovering in their inquiry about problem posing. (Note the essentially uncontrollable resurgence of self-referentiality.)

Analogies are made throughout the novel about the differences between mathematical and pedagogical situations and problems. One that is particularly worth exploring picks up on the often touted claim that in order for teachers to become effective in teaching problem solving, they must experience the same sort of problem solving as their students. While this may be a helpful condition for operating in a more exploratory mode in the classroom, it is just as important to think of teaching *itself* as a form of problem solving. The pedagogical issues, though integrated with mathematical concerns, outstrip a focus on the discipline. Such a point of view sees teaching not only as learning a craft, but establishes the value of living with uncertainty, debate, and partial and incomplete solutions. It ranges over domains such as the nature of pedagogical authority, the social context of the classroom, the purposes for educating, the multitextured relationship between mathematics and other fields of inquiry.

Options for coming to understand mathematics, pedagogy, and self are multiplied if characters in a novel can explore differences in perception and attitude regarding whether or not an alleged problem is in fact a problem, whether or not it makes sense, whether it is worthwhile or aesthetically appealing. It is helpful to have an array of opinions and points of view expressed by different characters so that students with a variety of interests and cognitive/emotional styles might find people with whom to identify.

Sy and Martha, for example, host a conference at their school and they invite five colleagues to present different and potentially incompatible points of view regarding the value of incorporating problem posing in the curriculum. Sy and Martha then ask the participants at the conference to evaluate the different presentations and to discuss their reactions further with someone who disagrees with their assessment.

Towards the end of the novella another controversy is presented, this time with Sy and Martha on the receiving end. Below is an excerpt in which they are called into the principal's office.

> In early December of the year that Sy and Martha submitted their proposal, Hyper Post was not spared the winter of all winters on the East Coast. Snow had begun to fall on November 12 and by the end of November, almost thirteen inches had already been recorded Sy and Martha were reminiscing in the teachers' room, thinking about the magnificent summer they had spent in Colorado and of the symbolism of the columbines that had covered Yankee Boy Basin. The semester had been a difficult one. Many of the parents became increasingly concerned that the open-ended and personal perspective that Sy and Martha offered would not get their kids into the best colleges, a point of view that Eloise was not reticent to echo. Though Sy and Martha were convinced that the concern was

not well grounded, it is true that their collaboration on the project had encouraged them to be even more experimental over the past couple of semesters. The bell for the end of the period sounded. As they were heading out to class, they heard an announcement over the loudspeaker from the principal's office. It was that deep, authoritarian tone that was more chilling than the frigid weather that they were to face in an hour. They were mentioned by name. The two of them were to report to the principal's office immediately. What was that tone again? It was the one that accompanied a similar request last spring for the two students suspected by most of the faculty to be heavily involved in drug dealing.

They walked into the principal's office, and he handed them each an envelope. The envelopes had been sent by special delivery from a government agency. The principal wanted them to open them immediately. "If there are charges being made against the two of you, for whatever reason, I need to know immediately. Please open it without further delay," he said. Martha thought she heard a laugh coming from inside the principal's private bathroom, and it seemed to be a familiar voice, but she couldn't identify it. Sy and Martha felt their privacy was still worth something regardless of the obvious intimidation. They left the office. The principal's face was still grim and they returned to the teachers' room. The name and address on each envelope had identical font styles. The three pieces of paper inside each of the envelopes appeared to be the same length and style. They did not quite have the courage to read the letters alone and so they agreed to alternate reading aloud. They decided to read the bottom page first. Sy read out loud. (pp. 82–83)

The package that is presented to them has come from one of the granting agencies to which they have applied for funding to write their novel. They soon find out that the package contains three letters from the NSF (National Science Foundation) reviewers. Below is an excerpt from one of them—a devastating letter—followed by Martha's reaction:

Dr. Jayne Courtmey, Director
National Science Foundation
Washington, DC

Dear Dr. Courtmey:

Thank you for selecting me to review the NSF proposal of Ms. Waters and Mr. Black. While I certainly found a number of their mathematical problems to be of more than passing interest, and while they express themselves well in writing, I cannot support their project. I fully subscribe to the view that mathematics is a problem solving activity. As a research mathematician I of course also appreciate the role of problem posing. Both of these activities however are a long time in the making. They are not the kind of activity that can be acquired in the absence of a great deal of knowledge and experience with mathematical ideas. What we need if we want to develop good problem solvers and posers is not explicit teaching of heuristics but rather good teaching of mathematical content at every grade level. We need teachers who can offer good explanations and we need teachers who know what a proof is and who can help students understand when it is necessary.

The use of a novel format is ludicrous because it obfuscates the most fundamental assumptions about the nature of mathematics. That is, there are very few fields in which it is possible to claim with certainty that something is true or false. Furthermore, there are very few fields in which the truth value can be either established or falsified by deductive proof. Mathematical creativity does not require that students explore issues from a personal point of view, but rather that they understand and create proofs and models that are valid and legitimate.

This proposal is based upon a total misunderstanding of the nature of mathematical thought, and I see no way in which I can suggest improvements. If the proposal is funded, I personally will do everything I can to see that the division within which you work will receive a thorough examination in the near future. (p. 83)

Martha was shaking. She hoped she could find some clue to identify the critic but such reviews were anonymous and she knew that if there were any clear identifying characteristics in the original letter they would be eliminated in the Xeroxed copy they received. Was there any reason to continue reading? Would the director have included that letter if the decision were positive? Sy persuaded her to read the middle letter out loud. (p. 84)

We leave Martha and Sy in the principal's office, waiting for the dénouement of their venture and of the novella. We turn now to a discussion of other novel adventures in education and in mathematics in particular.

There is some interesting precedent for the novel as an educational instrument in fields other than the obvious one of literature. Matthew Lipman has designed an entire program of philosophy for children—from the elementary grades through high school—around the creation and use of philosophical novels. (See Lipman, 1988, 1991; Lipman, Sharp, & Oscanyon, 1977; Lipman, Johnson & Gazzard, 1989). The program has had enormous success not only in affecting the philosophical thinking of children but in influencing their ability and interest in reading, in reasoning, in communicating and in problem solving more generally.

Part of what makes these novels enticing is that, though they deal with philosophical themes, they tend to do so in a relatively nontechnical way. In addition, they involve story lines that are robust enough to generate and reflect upon the philosophical issues. As in the vignette of Amy and Joanne, the story is not merely an "add on." In *Harry Stottlemeier's Discovery,* Matthew Lipman's first philosophical novel (Lipman, 1974), for example, a student's confusion of an assertion with its converse is pursued for a while without correction before other elements in the plot call it into question and it becomes an object of direct inquiry as an introduction to Aristotelian logic.

Such an orientation should be seen against a more general backdrop in philosophy that calls for a renewal of story-telling as a vital means of communication. Martha Nussbaum (1990) and Philip Hallie (1979, 1997) are well known philosophers who have begun to reexplore the power of stories and of dialogue, a return to the field's Platonic roots. Hallie (1997) who studied a pocket of goodness in the village of Le Chambon, France during the maelstrom of evil during the second world war, argues in quite personal terms for the value of story-telling in relationship to philosophical generalizations. Though articulated in personal terms, his point of view has clear educational implications. He says,

> At their best Kant's and Mill's philosophies are ingenious generalizations about particular people doing and feeling particular things. I can understand their principles only insofar as I can understand a story that embodies them. If there were no stories to illuminate their principles, I would not understand the principles at all. They would be words about words about words. Somehow, when my fellow (sic) philosophers put the rich accessible details of stories into their somewhat artificial languages, the stories lose much of their arterial vitality and much of their rich meaning. (p. 6)

Though Nussbaum (1990) also argues for the power of story-telling as a vehicle for communicating philosophy, it is clear, to the extent that she is concerned with understanding writ large, her comments are also part of a justification for using it in an educational format.

> How should one write, what words should one select, what forms and structures and organization, if one is pursuing understanding? . . . Sometimes this is taken to be a trivial and uninteresting question. I shall claim that it is not. Style itself makes its claims, expresses its own sense of what matters. Literary form is not separable from philosophical content, but is, itself, a part of content—an integral part, then, of the search for and the statement of truth. . . . But this suggests, too, that there may be some views of the world and how one should live in it—views, especially, that emphasize the world's surprising variety, its complexity and mysteriousness, its flawed and imperfect beauty—that cannot be fully and adequately stated in the language of conventional philosophical prose, a style remarkably flat and lacking in wonder—but only in a language and in forms themselves more complex, more allusive, more attentive to particulars. Not perhaps, either, in the expositional structure conventional to philosophy, which sets out to establish something and then does so, without surprise, without incident—but only in a form that itself implies that life contains significant surprises, that our task, as agents, is to live as good characters in a good story do, caring about what happens, resourcefully confronting each new thing. If these views are serious

candidates for truth, views that the search for truth ought to consider along its way, then it seems that this language and these forms ought to be included within philosophy. (pp. 3–4)

Egan (1988), more directly concerned with formal education, was drawn to story-telling as a teaching style in reaction to what he perceived to be the dominant mode of lesson planning by teachers, a mode that focuses on questionable/debatable ad hoc principles such as the following: teach from the simple to the complex, from the known to the unknown, from the concrete to the abstract. His brand of storytelling is inspired by standing on its head some of the empiricist assumptions about how one comes to experience the world. It is not that youngsters learn about good and evil by reading stories such as *Cinderella,* he argues. It is rather that they can appreciate these stories because they already have the concept of good and evil. He comments, "An impoverished empiricist view of science has misused the authority of science to promote in education a narrow kind of logical thinking at the expense of those forms of thinking which we see most clearly in children's imaginative activities. . . . As this essentially nineteenth century view of science finally recedes, and a more balanced view of science as a human activity becomes dominant, we in education might sensibly try to establish a more balanced view of children's thinking and learning" (p. 18).

Such a view of what one already knows and how one learns could enhance curriculum in mathematics significantly. We need not begin with the assumption that unless concrete realizations are carefully developed in texts beforehand, students will not understand the abstractions associated with them. Take the concept of number, for example. The assumption is probably wrong that we must proceed from natural numbers to fractions, to irrationals, to imaginaries on the grounds that students lack total appreciation for the concept of number until they have experienced the entire terrain. Suppose we based our thinking about the concept of extending number systems in a way that is analogous to Egan's thinking of "good and evil." That is, there are properties of number that students intuitively understand even before learning about them in a formal way.

There is some precedent in mathematics per se for taking dialogue and the storytelling format seriously. One thinks for example of Plato's *Meno* [Grube (1976)]; of excerpts from Hofstadter's (1979) *Gödel, Escher and Bach*; of some stories such as *The Little Dreamer* by Frederique Papy (1975) for young children; of the aforementioned *Proofs and Refutations* by Lakatos (1976); of the novel *The Rock: Surreal Numbers* by D. E. Knuth (1974); of the humorous short stories by Stephen Leacock

(1956) such as *Mathematics for Golfers*, of *Flatland* by Edward Abbott (1963)—the delightful story of life in a two-dimensional world.

The very titles indicated above say something about the possibility of a more vibrant genre—one that is more available in the elementary grades than in secondary education.[11] What sort of coherence and continuity is needed in order to have this genre become more than an occasional and supplementary source of stimulation? Some restraint would be needed to ensure that the mis-educative value of standard texts does not enter through the back door as qualities such as cohesion and continuity replace the more obvious forms of linearity and didactic format. What is needed is something that does not lose sight of one's purpose: to appeal to imagination; to relate mathematics to experiences that go beyond the pale concept of "application to the real world"; to find stimulation in controversy and paradox; to appreciate that one is as much engaged in meaning-making when entering a world of make-believe as when one operates in a world with concrete materials; to appreciate that acquiring an understanding of ideas is a fragile and idiosyncratic enterprise for which there are no universal pedagogical formulae.

While we might borrow from the insights of people such as Gay and Cole, Lipman, Egan, and others who have made such forays into the field of mathematics, it is worth thinking of the relationship of teaching to storytelling in more general terms. While the creation of mathematical novels would be a significant supplement to texts, it is worth keeping in mind that it is not only mathematical adventures that can be seen as story telling, but so can our lives as students and teachers interacting with each other in a classroom. We all have stories to tell that involve fears, conflicts, hopes, histories, projections for the future, insights about how our minds work, and insights about what it means to have a mind both within and beyond the context of mathematical thinking.

What kinds of stories and how they are told and exchanged is an issue that cannot be decided by a committee whose goal is to establish a national curriculum. What would seem to be required is a view of teaching as *revealing* rather than *hiding* self from others. It requires a view of teaching that is the antithesis of the macho and outmoded view of mathematics as something detached from self, totally deductive in spirit, and concerned primarily with generalizations of the widest possible sort. A significant agenda might begin with an educational impulse to connect curriculum with the deepest of human experiences and emotions of the sort that we portrayed throughout this book.

Problems and the Real World 215

As we have said, stories and storytelling are *au courant* concepts in much of educational discourse. They are seen not only as elements to reconstruct curriculum, but as tools for doing and reporting on research as well.[12] Although some outstanding mathematical stories portray some of the human qualities we described above, there are relatively few of them, and as far as I know, they have not been thought of as a centerpiece for curriculum reform. The concept of *novel* as text invites us to reconstruct our view of problem and of problem solving, connections with the real world as well as other themes we discussed throughout this book which—in the spirit of the Mitroff and Kilmann quote that launched this chapter—I think of as a story of sorts.[13]

Second Transformation: Secular Talmudic Format

As we have shown in our discussion of the different foci of the humanistic mathematics education movement (as in Table 1.1), we need to be cautious in making the divide too sharply. It is possible to think of the impetus for the concept of the novel as derived from a desire to transform mathematics teaching in order to focus on the *student as person*. A focus on *person as student*, however, suggests another transformation of text that is rich in its implications.

What other ways are there for us to view text so as to encourage an appreciation for the evolution of ideas? For the nonlinear qualities associated with their development? For controversy as a built-in element? For the sense in which ideas are integrated with and influenced by the culture within which they were born?

A not so modest answer is: The Talmud. The Talmud is a sacred text and considered to be second only to the Bible in Jewish tradition. It actually consists of two main texts: The Mishna (also transliterated from the Hebrew as Mishnah) and the Gemora (also transliterated as Gemarah). The Mishna was produced in the second century A. D. and is an attempt to codify traditions, especially in relation to the Bible. The Gemora, produced in the fifth and sixth centuries is commentary on the Mishna.

Rosen (1998, 2000), likening the Talmud to the internet comments,

> I have often thought, contemplating a page of Talmud, that it bears a certain uncanny resemblance to a home page on the Internet, where nothing is whole in itself but where icons and text-boxes are doorways through which visitors pass into an infinity of cross-referenced texts and conversations. Consider a page of Talmud. There are a few lines of Mishnah, the conversation the rabbis conducted (for some 500 years before writing it down) about a broad range of legalistic

questions stemming from the Bible but ranging into a host of other matters as well. Underneath those few lines begins the Gemarah, the conversation *later* rabbis had about the conversation *earlier* rabbis had in the Mishnah. (Rosen, 1998, pp. 8-9; emphasis in the original)

Both Mishna and Gemora deal not only with abstract principles of action but with the most detailed analysis of practical problems faced every day by members of the community. This observation about the Talmud, however, does not begin to describe its vibrant nature. It consists of a collection of stories, comments on secular and nonsecular behavior, aphorisms, and much more.

In fact, the Talmud is a first-rate example of the mixing of two traditions Bruner (1986) describes as *explanatory* and *interpretive*. The former is focused more closely upon precision and conveying accepted truths; the latter upon ambiguity, metaphor, and dialogue. Neither of the two stands alone. Each form provides a healthy backdrop for the other, and we tend to overstate the case when we pit for example, analytical thinking against a postmodern perspective. This is not a debate to be decided in an a priori way by the authority of educators, but rather is something each of us has to come to understand in idiosyncratic, personal, and perhaps situation specific terms.

One of the most interesting qualities of the Talmud is its nonlinear format in a global—but not a local—sense. It is assumed as soon as we begin to study the Talmud that we are already familiar with the entire collection—the ultimate in nonlinear thought. It is assumed that we all have experience in some sense with ideas that are to be refined through a variety of source material. As an indication of the nonlinearity of the experience, we are frequently urged to "recall" in an early section something that is first disclosed in a later one.

What makes it even more appropriate as a study of evolution of ideas is that each page not only has Talmudic text from the two periods, but that the text is surrounded by commentary of well known rabbinical scholars (Maimonides, Rashi), as well as anonymous critics whose reactions span hundreds of years. They explore the meanings and different interpretations of the Talmudic text, and they raise questions that are tangentially connected with it as well. Lukinsky (1987) speaks of its educational uses as follows:

> Modern books are linear and present the perspective of the individual author. The author's inner deliberation is for the most part buried. . . . If another view is presented at all it is usually refuted, . . . not given a fair presentation. . . .

> The reader is left to his own devices to put the argument together. . . . The framework and presuppositions of author are hidden in the editing, and they may seem to be speaking past each other. (p. 1843)

Much of the Talmud is problem oriented, i.e., problem solving oriented, but frequently a carefully reasoned debate with conflicting orientation ends with the statement: "Tie." That means that neither side has a definitive enough argument and that the reader is left to decide on the right conclusion or course of action.

In addition to being problem oriented, it is equally focused on problem generation, as we have been describing it throughout. Frequently it is in the effort to solve a problem that another problem is posed. Sometimes such posing is done in the service of solving that problem. Sometimes, however, it is not clear where and why a problem is posed in the first place.

Figure 6.2 is a sketch of a typical Talmudic page.[14] Though each page does have the Talmudic text as well as commentary, the style of each page changes depending on the particular sources of commentary and on other matters, such as the relevance of biblical sources. If this kind of activity sounds overly elitist, it is worth remembering that not just aspiring rabbis and learned people in the community, but all members are expected to engage in Talmudic study on a daily basis.[15] It is integrally connected with how one lives one's life. In addition, the study of Talmud is a collaborative act. It is studied in pairs so that each pupil can alternate the roles of teacher and student. This study may have been one of the earliest forms of cooperative learning.

It is impossible to study any page of the Talmud without hearing the voices of generations past and without feeling an intimate connection with those rich traditions and debates. As Rosen (1998) comments "Both the Mishnah and the Gemarah evolved orally over so many hundreds of years that even in a few lines of text, rabbis who lived generations apart give the appearance, both within those discrete passages as well as by juxtaposition on the page, of speaking directly to each other" (p. 9).

How does one begin to use the page of the Talmud as an element in a humanistic view of mathematics education? For many mathematical topics, there is a rich history that spans hundreds or even thousands of years. Though this history can be brought to bear in a way that reveals its evolution, one need not take the model of evolution as the primary one in order to borrow the Talmudic style. It is possible, for example, to have commentaries assume quite different educational formats.

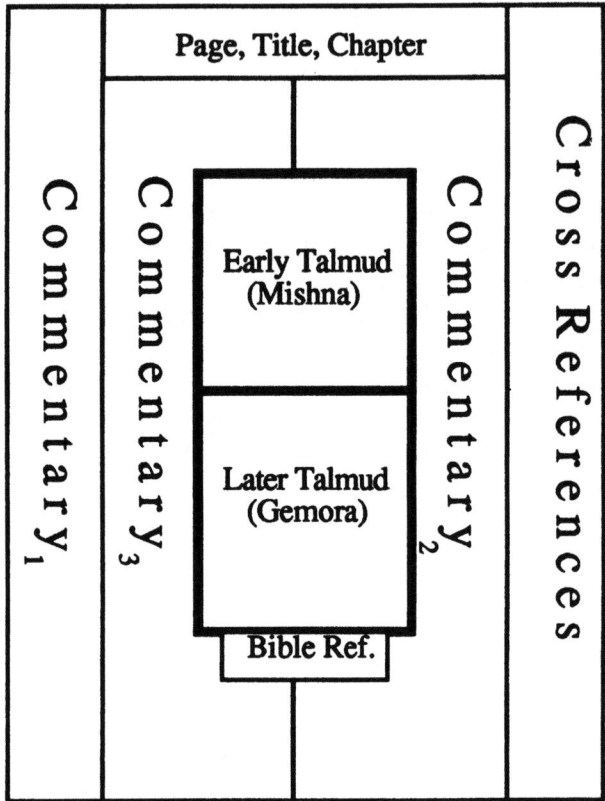

Figure 6.2 Schematic representation of a Talmudic page (adapted from Lukinsky, 1987).

Imagine, for example, that one of the commentaries on the page is the voice of the confused pupil for any idea that is developed. That voice can talk explicitly about why it is difficult to understand an idea and why it may in fact make no sense. Another voice may seek application of the idea to other mathematical and nonmathematical contexts. Another can explore the value of intentionally misunderstanding the idea, thus investigating the ways in which errors can be productive. Another voice might talk about how the particular problem reveals something interesting about the nature of mathematics. Yet another might persuade the student to find personal meaning in the debated topics.[16]

Part of what could be incorporated as Talmudic commentary is a voice that searches for what I described earlier in this chapter as "pseudo-history." This voice does not ask what actually happened, but what *might* have happened that created an interest in a particular topic, or what *might* have been an earlier rendition of an idea being studied, or what *might* have accounted for difficulties people had in understanding the idea. Some of these voices could be created by students as they attempt to unravel the meaning and significance of emerging ideas.

One interesting way to launch a Talmudic project might be to select among the many easily accessible (but difficult to resolve) paradoxes in mathematics (such as Zeno's paradox, or problem 1, the famous ski lift variation of Einstein's problem described in chapter 4) in order to find out how they were viewed over time.

Another might involve an implicitly central feature of secondary school mathematics: the extension of number systems. Earlier we spoke of the inappropriateness of assuming that students know nothing intuitively about fractions until they are explicitly taught them. A Talmudic approach in the history of mathematics could take each extension of the number system and create and relive debates of both a technical and philosophical nature. It might inquire into what it is about the concept of number that has made its expansions so difficult to handle over the centuries.

There are numerous ways in which a Talmudic format might be helpful in teaching from a humanistic tradition. In early stages of Talmudic courses I have taught, I usually introduced some simple phenomenon that appeared rather dull and uncontroversial. For example, I have asked students to discuss what they thought the distributive property was all about. For the most part, when they stipulated an algebraic rendition of the property, I asked them to think of the matter geometrically. I used this simple entrée as an invitation to compare what they "see" and what they find out by viewing the property from these two perspectives. When asked to create generalizations of the property from both a geometric and an algebraic point of view, they began to appreciate that there are connotations of each of the mathematical perspectives that are quite different. To generalize, for example, from the addition of joined rectangular regions to that of other shapes that are joined, leads to a consideration of algebraic formulations that are much richer than ever imagined in the case of rectangle.

Next, I invited them to come up with "real-world" counterparts of the property that are not strictly speaking mathematical or scientific in nature. For example, we began to explore how the concept of *love* might share (or fail to share) some important elements of that property. How does "loving my wife and my students" compare with "loving my wife" and "loving my students"? Does "love" in both cases function in the same way that multiplication functioned in the algebraic rendition of the property?

What new insights are raised about the meaning of the property in a mathematical sense and in a "real-world sense" when such applications are unpacked?[17]

Eventually, we started writing some of this down in Talmudic format. I began the Mishna either by asserting "given" and non-controversial facts or by introducing two points of view that appeared to be in tension with each other. I then followed it up with Gemora commentary in which the issue or the controversy is further revealed by a variety of voices.

At the beginning of the course, students generally felt more comfortable completing some Mishnaic issue that I had created than creating one on their own. Eventually though, they created their own Talmudic text to work through. Perhaps the most interesting phase of the course occurred when students requested that they cooperate with each other by creating Talmudic commentary on each other's work. In one class, students spontaneously decided not only to create Talmudic format for the mathematical/pedagogical issues that interested them, but they chose to devise a weekly diary in which they used the Talmudic format to discuss their reaction to the course itself.

In order to convey how substantive issues might be portrayed in a Talmudic format, I have included a one-page course disclosure of a course on the Talmud for secular purposes that I taught to secondary school mathematics teachers (Appendix A). The disclosure is designed according to the very format that will be studied in the course itself (a point of view that should have a familiar ring to the reader by now). The Mishna speaks matter of factly about the nature of the course, almost as if it were merely descriptive and nonproblematic. Following the introduction, the Gemora begins to question and conjecture what is meant by the components that comprise the introduction. Based upon these comments, new insights are expressed in the margin of the page.

For the purpose of suggesting something about the manner in which mathematical pedagogy can be developed, I provide in Appendix B a Talmudic excerpt of a topic that we discussed in chapter 4 on "same/different." In particular I focus upon the concept of isomorphism by using the simple example of adding numbers that end in 0. The Mishna suggests that adding two numbers ending in 0 is a piece of cake, but points out that the identical procedure does not work for numbers ending in 5. A variety of explanations follow in the Gemora.

It then reflects upon the concept of "short-cut" itself, a concept that is frequently invoked in mathematical thinking, but that is not well analyzed. Thus, previous "background" material of little consequence shifts to the "foreground" of the conversation. Bringing to the foreground something that was earlier assumed as unproblematic not only honors possible concerns of neophytes who are plagued by a forest and tree problem on the one hand. On the other hand, it has the potential to open up new territory for investigation. Switching background and foreground is in fact a disguised form of What-If-Not.

After working through a number of different possible explanations for why the short-cut works, the Gemora stops its logical development (explanatory mode) and "out of the blue" begins to talk of a dream of one of the narrators whose daughter is pregnant (interpretive mode). He wonders about the gender of his unborn grandchild. He is then whisked away in a chariot and visits a different land that is in every way like his own, except for the fact that time and space are contracted. Everyone is considerably smaller, and events are played out in a fraction of the time it would take in his "real world." Without amniocentesis, he is thus able to discover in a few seconds the gender of his unborn grandchild.

As the text continues, commentary in the margin begins to explore the connections between the rather simpleminded short-cut of adding numbers ending in zero and the narrator's concern over the gender of his forthcoming grandchild. Eventually, a connection is made between the "real world" of the story and the arithmetic short-cut. The dream world is talking about a structure that is analogous to the one used in arithmetic. Determining the gender of his grandchild months before it is born has the same structure as the arithmetic short-cut. Thus, the concept of isomorphic structures is explored and revealed without fanfare and without making use of technical machinery that usually introduces and obfuscates what that concept is all about.

When teaching Talmudic courses, I have found that this format not only enabled me to incorporate many of the elements about problem and real-world connections in a humanistic tradition, but it has had a personal impact on my students that was far more powerful than I had expected. Many students commented that even when we were not explicitly making connections between their lives and what we were exploring, they ended up seeing themselves as students, teachers, friends, spouses, or parents in a different light. I am still trying to understand why these connections are made and how they account for such strong personal feelings.

We end this section by commenting on one interesting Talmudic starting point that was an eye-opener for virtually all of the students: Euclidean geometry as Euclid himself appeared to have viewed it. His understanding of the relationship of length to line segments and of area to regions in the plane is unbelievably mind-boggling (as uncovered by reading sections of Euclid's *Elements*). It comes as a surprise, even a shock, to most people to discover that first of all, he had not a single formula for areas of figures in the plane, e.g., the area of a triangle equals one half base times height, and that second he was able nevertheless to discuss the concept of "same area." For example, he showed in the Pythagorean theorem that the sum of the areas of the squares on the sides of a right triangle equals the area on the hypotenuse without reference to any formulas for the areas of geometric figures. He did so by engaging in the child-like act of cutting and pasting. Similarly, because he was not able to handle incommensurable line segments, he was not able to make sense of the concept of *the ratio* of two line segments. Even so, he was able to tell us when *two* ratios—each of which made no sense—were the same!

Euclidean geometry of course leads us to consider non-Euclidean geometry as a Talmudic text source. This resource might well explore the centuries old controversy that called into question the parallel postulate, a topic we discussed in chapter 3. What is particularly appropriate about that topic, in light of many of the concerns of this book, is that it single-handedly not only solved a problem, but overturned the perception of geometry as a field that is based upon "self-evident" truths.

A central player in the creation of non-Euclidean geometry was, as we mentioned, the Hungarian mathematician, Janos Bolyai (1802–1860). He had attached himself not only to a *problem* but to a *program* that was fraught with uncertainty, one that had the potential to raise the most fundamental questions about the nature of thinking in mathematics and beyond. It was a problem that had become addictive to him, however, and despite the fact that factions had formed which were taking a jaundiced view about the value of pursuing the problem, he persisted. As (an only slightly contrived) grand finale to this section, chapter and book, we recall the advice that his father, also a mathematician gave to him at the time: "For God's sake, I beseech you, give it up. Fear it no less than sensual passions because it, too, may take all your time, and deprive you of your health, peace of mind, and happiness in life" (Boyer, 1968; p. 589). So much for mathematics as purely logical, ahistorical, unconnected with reality and detached from human drama. So much for education as passing on an hermetically sealed version of the wisdom of the ages.

APPENDIXES

Appendix A

Talmudic Style Course Disclosure

The Talmud in Secular Education

Mishna

As it is said, the course will be about *itself*. It is open to students—past and present—in all fields who are interested in experiencing a modern secular transformation of a two–thousand–year–old religious tradition. We will explore and elaborate upon a form of critical and creative thought that is deeply personal and of the world. The focus will be upon a variety of educational roles and issues and will depend in part upon students' emerging interests.

Gemora

Reb Stevenson asks: What can it mean—that something will be about itself? As the sages taught us, for a thing to be about itself means that it *exemplifies* itself in some way. The course's action then will exemplify the very structure it is trying to study. What is this structure? He suggests that perhaps the layout of this very page will reveal something about that structure.[1,2]

Reb Weis comments, "As it is said," what is the significance of that remark? It appears that we must be participating in an ongoing conversation, akin to some postfeminist dialogue that was not made explicit.[3]

Reb Seller asks: What does it mean that it is open to past and present students? That it refers to past students means that the course was offered in the past, but that former students would find the new course to be an extension and elaboration rather than a repetition of the past offering.[4,5]

Reb Schroeder points out this Mishna is silent on the readings and course organization. He comments that this experience mirrors what he recalls from the Mishna entitled "Brown chutzpah."[6]

[1] Reb Hosenfeld points out that the page itself is in dialogical form. It can be read as a form of conversation.

[2] Reb Finn comments that the conversation may be taking place over a protracted period of time. Perhaps the students will record and revisit their commitments to some earlier classroom dialogue on later occasions.

[3] Reb Dimitriadis comments that we may be seeking to locate the place of non-linear and associative thinking as a culturally significant element not only of teaching and learning, but of research as well.

[4] Reb Nyberg tells us that in the past the course was offered with a mathematical focus. Now it appears to have an orientation that is more epistemological.

[5] Reb Prinzivalli suggests that those who had a previous Talmudic experience with this instructor might, in the spirit of cooperative learning, wish to assist in the orientation of new students.

[6] Reb Volfson surmises that there will be philosophical readings and excerpts from literature; also the students will communicate with each other in Talmudic format.

Appendix B

Excerpts from a Mathematical Talmud

Sameness

Mishna

The House of Shamai and The House of Hillel return from purchasing corn seeds. They promise to pay the corn seed merchant in a month with the money they will receive from the father-in-law of The House of Shamai. Together they will be building a new barn for him, and he has promised to pay their total expenses for their corn seed as soon as they have framed the structure. The corn seed merchant has told The House of Shamai and The House of Hillel that he has figured out how much they owe based upon the information they have given him for each of the two plots of land they will be sowing. He has said that The House of Shamai owes 470 shekels and The House of Hillel owes 220 shekels. They each calculate separately the amount of money they will have to request from the father-in-law of The House of Shamai.

The House of Hillel does his calculation in the sand and figures out that it will be 690 shekels all together. He did the addition as follows:

$$\begin{array}{r} 470 \\ + 220 \\ \hline 690 \end{array}$$

The House of Shamai also figured out the sum and it is indeed 690, but it is possible to do so without moving around so much sand on the ground. He says that it should be done as follows:

$$\begin{array}{r} 47 \\ + 22 \\ \hline 69 \end{array}$$

And then The House of Shamai says that you should just tag on a zero at the end and make the answer 690.

The House of Hillel says that this is a trick which seems to work in this case, but that it is not the way to do addition. He asks: If the sum was 475 and 225 shekels would you so the same thing? He shows the two procedures:

$$\begin{array}{r} 475 \\ +\ 225 \\ \hline 700 \end{array}$$

$$\begin{array}{r} 47 \\ +\ 22 \\ \hline 69 \end{array}$$

Then if you tag along the common 5, the answer would have to be 695, when you know it is really 700.

Gemora

Reb David tries many examples of adding pairs of numbers each of which ends in 0. Indeed the procedure of The House of Shamai—ignoring the zeros for each pair at the beginning and then appending a zero at the end—works in all cases he has tried. He believes it will work always.

Reb Li thinks about the matter and realizes that there is a simple reason that the procedure works with zeros at the end of each pair. She observes that if you follow the standard algorithms for adding numbers, then if you start at the most right-hand corner of each pair and if each member of the pair ends in 0, then you are just adding 0 + 0 and of course you end up with a 0 as the sum.

Reb Markus claims that in order to make sense out of the procedure, it is necessary to appreciate what the little zero is doing at the end of each of the numbers. It is really not just adding a little zero to each of the numbers of the original pair. What, for example, is the relationship between 47 and 470? What is the relationship between 22 and 220? He points out that 470 is really ten times 47; also 220 is really ten times 22. When you look at the sum, 690 is really ten times 69. Thus, what we really have is the following:

$$47 \times 10$$
$$+22 \times 10$$
$$\overline{69 \times 10}{}^1$$

Reb Fendel points out that Reb Markus is not wrong, but not everything is fitting into place. For one thing the problem does not begin with finding the sum of 47 and 22, but rather the sum of 470 and 220. He knows that what Reb Markus has done is relevant, but despite the fact that everything looks so simple, he wonders if there might be something beneath the surface that is missing. He is tired and lies down to rest in the field next to his farm. As he lies down, the last thing he recalls is the good news he just heard from his daughter—that she is in her first trimester of pregnancy. He has the following dream.

He dreams about his forthcoming grandchild and wonders if it will be a boy or a girl. If it is a girl, his son-in-law will have to think about how to prepare a proper dowry in about twenty years. Reb Fendel wonders if he himself might be able to contribute to that dowry by passing along some of his land that he imagines will be rich in corn in years to come. He should only live so long. If it is a son, he imagines that the boy will accompany both his son-in-law and himself to services every Sabbath. He should only live so long. It crosses his mind that he might also like to have a granddaughter accompany them to services, but no one in his shtetl ever heard of such a thing. Of course, he then smiles as he imagines that one day, men might have to come to the family of their prospective wives with a dowry. He is imagining what family members the child might

---

[1] Reb Vilson points out that Reb David, Reb Li and Reb Markus are each looking at proof in a different way. One is trying to establish the generalization based on an empirical mode by trying many cases; one is coming up with a proof that works for all such cases. Such a proof, however, makes use of an algorithm or a procedure, and though it proves something, it does have a blind quality to it—unless one already understands the positional notation on which the algorithm depends. The third proof gets at the meaning of the notation, at least in part. Reb Vilson wonders if there are some other proofs that might be even more revealing. He wonders what kinds of things a proof might be expected to reveal in general, beyond its ability to show that something can be shown to be true.

[2] Reb Fisher looks again at the three different proofs and wonders what role zero is playing even apart from the proof. She has heard people say that zero is not a number but is a place holder. She wonders which of these two conceptions is being revealed through the notations being used.

[3] Reb Putzik begins to think of something she had not thought about before. Her thinking is as follows. It is one thing to be able to understand the meaning of the positional notation used in numbers and even to be able to prove things with such notation. What she wonders, however, is why anyone ever was motivated to come up with the idea of positional notation. How might we have represented numbers before such notation was developed?

resemble, what the first words might be, when suddenly he sees a brightly colored horse-drawn carriage and an old woman dressed with emerald beads on her blouse and skirt. She beckons him to step up to the seat next to her on the carriage. He is shy and does not feel comfortable sitting next to a strange woman, but he steps aboard as she whispers that she has mysteries to reveal.

They chat away about the weather for a while when suddenly, the horse begins to take flight, hovering only a few feet above the ground at first, but eventually he can just barely see his farm from the air. After a seemingly interminable amount of time, he begins to see stars from a distance. They are all covered with a thin veil of tightly crocheted yellow lacy curtains — the sort his wife had made for his daughter when she was just a baby, in order to protect her carriage from flies as the family walked to town in the summer. He hears voices of his daughter and son-in-law—barely audible but still coming from the direction of his hometown. They are telling him that he is in for an adventure and that he should go with the flow and not resist, the way he resisted the last time he got a massage and was supposed to try to relax. He begins to fear that he will never return. He then hears his daughter and son-in-law singing, but their voices are no longer coming from his farm. They seem to be coming from a strange land that is miles and miles away from his village. The horse stops abruptly and lands on a piece of earth. At first it looks like he has returned to his own village. He sees the farm house in the distance, the plot of land on which he will plant corn, and he sees several people coming towards him from the distance. He embraces his wife, his daughter, his son-in-law. They seem to be the same as when he left on his horse-drawn adventure. They have the same clothing; they have the same smells; they talk the same; but something is peculiar. As he is hugging them, he realizes that he is lifting them each up into the air. They are about four feet above the ground. Why is that? Suddenly he realizes that, though they all have the same features, they are about one tenth their normal size.

His daughter—whom he is hugging—hits the ground with a thump and he realizes that he too has begun to shrink in size. His wife has prepared a picnic of his favorite gefilte fish and matzos. They toast his joyful return with some wine used only for special occasions. Reb Fendel wipes the dripping wine from his lips with his shirt sleeve and prepares to tell them about his adventure, when suddenly his wife calls him in from the field for dinner. He is just completing dinner, when she tells him to get ready to eat breakfast before going to visit her family in the next village.

Reb Fendel goes to Reb Yisroel, the village wise man. He tells him his dream and asks him to interpret it. Reb Yisroel asks Reb Fendel what he was doing before he began to dream. Reb Fendel tells him that he was busy trying to see in a clear light why it was that the sum of 470 and 220 could be calculated in such a way that the zero's were first ignored and then imported at the end. It was an amazingly simple idea he thought, but he wondered if there might be more to it than appeared on the surface. After some questioning, Reb Yisroel confirms that Reb Fendel is indeed obsessed with the question of what the sex will be of his forthcoming grandchild.

Reb Yisroel seeks a connection between the mathematical event and the dream. In both cases, he points out, you are trying to short-circuit an event. In the case of arithmetic, you are trying to ignore part of the calculation, the zero at the end, and you engage in a slightly simpler problem. You then tag the zero on at the end. It is not a monumental short-cut, but it is a short-cut nevertheless. A lot of arithmetic is about finding shortcuts.[5]

What is the short-cut in the case of Reb Fendel's grandchild he asks? Obviously Reb Fendel would be happy to have some procedure that would enable him to predict the future. He desires a way of knowing without waiting. If in Reb Fendel's make-believe world, time shrinks in such a way that one day is one minute, then what would normally take nine months or so (as in the gestation period of human beings) to find out, could be accomplished in 270 minutes or about four and a half hours. Reb Fendel would then not only know the gender of his grandchild, but he would actually be able to see a miniature version in just a few hours. Reb Fendel is happy and goes about his business for the rest of the day, though he does have unusual hunger pangs for the rest of the day and for the next few days as well.

[5] Reb Shmuel wonders what the different meanings of short-cut are. Sometimes it means that for some subset of a set of problems you can devise a procedure involving less work than you need to perform for problems of that set in general. He thinks for example of the wonderful short-cut he figured out for squaring numbers ending in 5. He realized that the answer always has to end in 25 and that . . . (he forgot the rest). Sometimes, short-cuts for a procedure are more efficient than an original procedure in all cases, not just for a subset of the original one. The hand-held calculator may be an example of a short-cut device. But a short-cut is always a short-cut in relation to an alternative way of operating. Nothing is a short-cut unless it is a short-cut in comparison to some already accepted procedure. So something is always a short-cut regarding a specific procedure in relation to something else. Nothing is "just a shortcut." This does not sound like a great realization to Reb Shmuel, but he sees that he may have hit upon a way of thinking about things that has wider applicability. It is the first time he realized that when he claims that "any Procedure X (a new algorithm) has the property of Y (being a short-cut in this case)," there may be a hidden qualifier. That is, X has property Y in relation to Z (an already known algorithm). He will look for other applications of this concept.

Reb Tamara thinks some more about the connection between the mathematics and new world that Reb Fendel experienced in his dream. She notices that the dream has a lot built into it other than the ability to predict the future because of the short-circuiting of time. Though it may seem obvious, she decides that what is obvious is sometimes most easily overlooked. Here's what she sees:

1. There is a correspondence between the two worlds of Reb Fendel. For every person in his real world, there is a facsimile in the dream world. This is like the concept of the alter ego that he has heard several people speak about. They imagine that somewhere in the universe there is someone who looks and acts just like them.

2. Though Reb Fendel and Reb Yisroel focused on pregnancy, it might be worth moving back one step. Since it takes "two to tangle," look at one stage before pregnancy. That is, assume that you are interested in the "operation" of mating for two people. Then when a male and a female mate, they will (with some luck, or perhaps the opposite in some cases) produce an offspring in nine months or so.

3. What is most interesting from Reb Fendel's dream is that he can translate mating of his daughter and son-in-law in the real world into their mating in the imaginary one.

4. The advantage is that the offspring born in the imaginary world, having a correspondence with one to be born nine months later in the real world, can be seen in some form almost instantaneously despite the fact that its realization will take nine months longer in the real world.

Reb David points out that what is of most interest in the scheme of Reb Tamara's, is that because of this correspondence between the two worlds, you have the following shift of focus:

1. Focus first on the elements of the real world (pairs of people) who perform some act (mate).

2. Then look at their alter egos in the dream world and at the process of performing some act there as well (also called mating).

3. Next is the result, which is quickly determined (in relation to time in the real world).

4. Then find the alter ego of the new offspring in the real world in order to make a prediction about the sex of the yet unborn offspring.

Reb Cooney points out that the kind of correspondence between the two worlds is an interesting one. It is assumed not only that you can find an alter ego, but that there is exactly one element from the real world that corresponds with one in the imaginary world and vice versa.[6]

Reb Fradin is a bit confused. She decides that a picture would help convey the dream-world / real-world phenomenon of Reb Fendel. Here is what she comes up with.

[Talmud in Progress: left to the reader to complete]

---

[6] Reb Rising points out that this is a big step. It is establishing not only that for every element in one system there is another one in the other system. Every element in one system could have more than one element in the other. Here a one-to-one correspondence is assumed. For every element in system A, there is exactly one in system B and vice versa. In mathematics, if we look at the relationship of pairs (x,y) generated by the equation $y = x^2$, then for every element in the x domain, there is exactly one in the y domain, but not vice versa. He thinks a bit more about the conditions under which there is one-to-one correspondence versus other alternatives.

Notes

Preface

1. See Brown (1996a) for the upshot, *Posing Mathematically*. It is discussed further in chapter 6.
2. Though tied (sometimes implicitly) more closely to problem solving and less to personal understanding than is the case in much of the analysis in this text, there has been a small pocket of excellent essays that illuminate the many faces of problem posing. See for example Burton (1986b); Kilpatrick (1987); Silver (1994); Silver, Mamona-Downs, Leung, & Kenney (1996). Many of the essays in the edited collection by Brown & Walter (1993) demonstrate how colleagues have used problem posing (and especially the What-If-Not scheme developed in this text) explicitly and powerfully in their teaching of problem solving. The works of Butts (1980) and Stein (1996) shed light on a topic that has become popular recently—problem posing more as an activity for teachers than for students. Here the focus is on teachers making use of strategies to pose problems for students so as to better entice inquiry. Frequently these strategies are designed to encourage more open-ended exploration than would otherwise be the case. Butts (1980), for example, presents us with three different formulations of a problem, each making a different assumption regarding the sort of enticement that flows from the way in which a problem is formulated:

   Problem 1. Let d(n) denote the number of positive divisors of the integer, n. Prove that d(n) is odd if and only if n is a square.

   Problem 2. Which positive integers have an odd number of factors? (Justify your answer).

   Problem 3. Imagine n lockers, all closed, and n men. Suppose the first man goes along and opens every locker. Then the second man goes along and closes every locker beginning with #2. The third man then goes along and changes the state of every third locker beginning with # 3 (i.e., if it's open he closes it, and vice versa). If this procedure is continued until all n men have passed by all the lockers, which lockers are then open? (p. 23).

3. Bliuma Zeigarnik, a psychologist, did some interesting research on how the extent of resolution affects our ability to remember and be haunted by ideas. She showed that ideas that are left unresolved, as in reappearing in several forms of a conversation but not highlighted by definition or summary, are more prone to be remembered than those that have been "resolved." Her name has become a common noun to describe her idea. See Zeigarnik (1965).

4. As a matter of fact, the manuscript for this book was completed and submitted before the publication of the fourth *Standards* document. For a listing of the documents, see note 3 of chapter 1.

Chapter 1 Reform in School Mathematics: A Backdrop and Critical Overview

1. As will be seen in chapter 6, this phrase influenced *The Standards* as well, though it had a different connotation.

2. See for example, Kline (1966a). In Brown (1968), I summarized such dissenting voices according to the following categories: the relationship of mathematics to intuition, creation of mathematical ideas, the relationship of mathematics to science, and goals of teaching mathematics.

3. Three *Standards* documents, focusing on curriculum, teacher education, and assessment, respectively, were all published by the National Council of Teachers of Mathematics between 1989 and 1995. They are entitled: *Curriculum and Evaluation Standards for School Mathematics,* (1989), *Professional Standards for Teaching Mathematics* (1991), *Assessment Standards for School Mathematics* (1995). A fourth document, *Principles and Standards for School Mathematics* (2000), combines the foundation of the first document with the other two. Among national reports that were produced by scientists and mathematicians are: National Research Council (1989). *Everybody Counts: A Report to the Nation on the Future of Mathematics Education*, and Steen (1990a). While I am selecting commentary mostly from the perspective of the U.S. educational system, there is a great deal of reform on the international scene as well. For a discussion of such commonalties, as well as significant differences, see Bishop et al. (1996). In particular, Comiti and Ball (1996) speak of teacher preparation from a comparative perspective. It is important to keep in mind that my purpose is less to evaluate *The Standards* movement (nationally or internationally) and more to forge a new paradigm that has been acknowledged only marginally.

4. To some extent, the concept of humanistic education can be seen as a reaction to a view of liberal education, one that claims to choose its disciplines based largely upon distinguishing characteristics among the many different fields of human inquiry. Though the particular fields have changed over time, the characteristics that have been picked out for over two thousand years are ones that separate a field from others by virtue of special qualities of the field—qualities that have minimal overlap with others. See Kimball (1986) for an analysis of the many different conceptions of liberal education. See Hirst (1970), Hirst (1974) and

Phenix (1964) for a justification of the centrality of the disciplines in liberal education. See also Gardner (1999b) for an excellent analysis of the centrality of the scholarly disciplines in contributing to a deep understanding of truth, beauty, and morality. See Brown (1986a) and Martin (1981) for a tempering of these points of view. Some of that tempering is discussed in the sections that follow and in much of this text. The view of humanistic mathematics education that we portray draws upon the disciplines, but integrates them within a broader conception of personhood.

5. A modified form of this story appears in Brown (1981c). Several other portions of this section appeared in modified form in Brown (1996b).

6. See Scheffler (1973) for a discussion of this issue under the concept of educational relevance.

7. Two relatively non-technical accounts of his proof and of the power of self-referentiality—the driving force in his argument—may be of interest. One of them, Nagel & Newman, 1958, focuses exclusively on mathematics as a discipline. The other, Hofstadter, 1979, points out the centrality of that kind of thinking in art and music as well as in mathematics. See Dawson (1999) for a discussion of Gödel's genius and his psychological make-up.

8. Ernest also edits an international newsletter entitled *Philosophy of Mathematics Education* (POME), which explores many of the issues raised in a program which focuses upon challenges to the absolutist schools of mathematical philosophy. The journal can be accessed from the web site http://www.ex.ac.uk/ ~PErnest.

9. Dewey in *The Child and the Curriculum* (1902) argues for the sense in which the popularized dichotomy drawn between progressive and traditional camps is a foolish line to draw. He coordinates the two points of view by seeing them as representing different stages in an evolutionary process rather than as points of view that are in basic conflict. Dewey comments, "What then is the problem? it is just to get rid of the prejudicial notion that there is some gap in kind (as distinct from degree) between the child's experience and the various forms of subject matter that make up the course of study" (p. 344).

10. For Dewey, an interest in the social context of learning was not so much a pedagogical ploy as it was a realization that all subject matter derives ultimately from roots that are social in nature. Dewey (1975) comments, "The moment mathematical study is severed from the place which it occupies with reference to use in social life, it becomes unduly abstract. . . . It is presented as a matter of technical relations and formulae. . . ." (p. 41). Here he was not so much calling for practical applications as we seem to be doing nowadays as he was inviting us to consider the ways in which people influence each other in the genesis of their thinking and in their motivation for defining areas of inquiry.

11. See White (1993) for a compilation of essays that portray the span of issues and themes in this journal. As of this writing, the journal has attracted an international membership of over two thousand subscribers.

Chapter 2 Problem Solving and Its Emerging Companion: Philosophical and Educational Roots of Problem Posing

1. Marion Walter and I published our first article about the connection between problem posing and solving in 1969. We eventually elaborated upon the theme in *The Art of Problem Posing,* first published in 1983, and we subsequently compiled writings of our colleagues that derived from our inquiry in *Problem Posing: Elaborations and Applications* (Walter & Brown, 1969; Brown & Walter, 1983, 1986, 1990). My novella (Brown, 1996a), written for teachers, describes the experience of two classroom teachers who embarked on a problem solving adventure in their teaching. It is mentioned in the preface and described further in chapter 6.

2. We have presented the postulates in a slightly more readable form than is the case in *Euclid's Elements.* Euclid's precise formulation in Heath, 1956 is:

    Postulate 1: "Let the following be postulated: to draw a straight line from any point to any point" (p. 195).
    Postulate 2: "To produce a finite straight line continuously in a straight line" (p. 196).

3. Even though we have revised the statements of postulates 1, 2, and 5 for the purpose of enhancing readability, the slightly revised reformulations of Euclid have an interesting dynamic quality that is lacking in modern-day formulations. Thus compare the two following renditions of postulates 1 and 5:

    Postulate 1: "A straight line can be drawn from any point to any other point" (Euclidean form).
    "There is a straight line between any two points" (modern form).
    Postulate 5: "Through a given point not on a line, exactly one line can be drawn that is parallel to the given line" (Euclidean form).
    "Through a given point not on a line, there is exactly one line parallel to the given line" (modern form).

    In the modified Euclidean rendition for each case, there is a sense of human agency. In modern-day renditions, there is merely an existential remark (indicating that something *exists,* but what is left out is that it is created by a human act of "drawing" something). This point was brought to my attention by John Corcoran, a philosopher of logic at the State University of New York at Buffalo. We can imagine that the difference may be psychologically quite powerful.

4. It was shown that a system that accepted the other (more primitive) postulates but included the negation of the parallel postulate was itself a consistent system. This is an essential strategy for proving that a statement is independent of others.

5. See, for example, Laudan (1977, 1984), Martin (1988) and Scheffler (1967) for a discussion of issues related to the distinction between discovery and verification. Feminist epistemology has begun to seek philosophical understanding of the concept of discovery, which was heretofore relegated to the field of psychology.

6. As far as I know, Marion Walter and I coined the expression to describe a strategy as well as an orientation towards problem posing. The expression first appeared in our publications in the late 1960s. See Brown & Walter (1993) for a collection of essays written by colleagues who have adapted and expanded upon the strategy.

7. There has been some interesting research in teacher education since the mid-1980s that has been focusing on the theme of "noticing." Much of it is based upon the research agenda of John Mason (1994) from England as well as his disciples. A heated discussion of some of the issues as well as of the overall concerns of that program can be found in the *Philosophy of Mathematics Education Newsletter # 8*, edited by Paul Ernest (1995). See note 8 of chapter 1 for web site.

8. Here and in the next two sections, we are seeking connections between a What-If-Not perspective and the concept of understanding. For a more full blown philosophical analysis of the concept of understanding as it relates to education, see Martin (1970) and Scheffler (1965). For a philosophical analysis that relates specifically to mathematics education, see Brown (1976), Lehman (1977), Poincaré (1961) and Sierpinska (1990). A valuable beginning is to appreciate that it is an ellipsis to claim that someone understands anything. To say, for example, that one understands that in a right triangle, $x^2 + y^2 = z^2$, what is being asserted? It can be intended to imply that someone understands what that statement means. Such an understanding would not necessarily imply that someone can prove the statement. It could mean that someone knows how to prove the statement but cannot necessarily apply it to new situations. Further, we are often hopeful that students who understand in one sense would be able to "transfer" their knowledge to new and different circumstances, and that is of course a relevant concern. Some interesting research has been done in that domain and also much confusion has been generated in the conception and interpretation of that research. Many interpretations of research on "situated learning," for example, arrive at the erroneous conclusion that knowledge does not transfer except under extraordinary circumstances. Anderson, Reder, & Simon (1996) take Lave (1988) to task for making such claims as:

> Place-holding algorithms do not transfer from school to everyday situations, on the whole. On the other hand, extraordinary successful arithmetic activity takes place in these chore (shopping, selling produce, making and selling clothes) and job settings. (p. 149)

> It is puzzling that learning transfer has lasted for so long as a key conceptual bridge without critical challenge. The lack of stable, robust results in learning transfer experiments as well as accumulating evidence from cross-situational research on everyday practice, raises a number of questions about the assumptions on which transfer theory is based. (p. 19)

They point out not only that Lave's claims are logically inverted (since she has demonstrated not that school learning does not transfer to real life situations, but rather the other way around) and that the general skepticism about transfer is based upon the inclination to look for transfer where one is least likely to find it.

9. See Scheffler (1960) for an analysis of the concept of definition and its educational uses. The mathematics education community seems to have truncated the range of meanings and uses of "definition" in its educational agenda.

10. The issue is not only a pedagogical or psychological one. Beneath such concerns lurks a form of Plato's famous dilemma of Meno (see Grube, 1976, & Petrie, 1981). The dilemma concerns a paradox regarding the nature of inquiry. The question is: How is it logically possible to ever come to know anything. On the one hand, if we already know it, then no inquiry is necessary. On the other, if we do not know it, then how can we ever inquire since we do not know the very thing about which we are to inquire.

Chapter 3 The Concept of Problem and Its Educational Fallacy

1. To point out Dewey's inclination to collapse "perplexity" and "problem" and to join problem with solution is not meant to trivialize his more general program of analyzing what is involved in the activity of reflective thought. He is painstakingly clear in pointing out that there is no formula that will enable us to arrive at solutions. Furthermore, he observes that we tend to find it quite difficult to suspend judgment—an essential quality of reflective thinking.

2. It is interesting to recall one problem that Schlick selects as impossible in *practice* but not in *principle* to answer. He mentions that questions about the moon fit the first category because "It is . . . technically impossible for human beings to reach the moon and go around it . . . but we cannot declare it impossible in principle" (p. 25). The fact that he was wrong about the technical possibility of exploring the moon of course does not weaken his argument. It actually strengthens his point that we need not focus on practical possibilities in order to establish the logical status of a question.

Chapter 4 Problem Posing/Solving in a Humanistic Light: Softening the Fallacy

1. See Brown (1996b), p. 1304 ff. for a further elaboration of this anecdote.

2. The focus on scope and intentionality in this section has been on the educational goals of problem solving from the perspective of the individual. See Jarvie (1979) for a discussion of competing goals of problem solving in science. He explores the thesis that competing theories are to be judged by their potential to solve the greatest number of empirical problems.

3. Brownell (1935, 1945) was an early advocate of connecting learning of mathematics with meaning. He juxtaposes meaningful learning with rote learning—as we do today. Based upon the fundamental idea that number is an abstraction and that "arithmetic when viewed as a system of quantitative thinking is probably the most complicated subject children face in the elementary school," he proposes a

variety of experiences for students to come to understand what a number is, how numbers relate to each other, and how and why they can be combined with each other. Furthermore, Brownell appreciates that such discoveries are not made once and for all, but that they have to be reexperienced in a variety of contexts over a protracted period of time. Most importantly, he showed that even when "meaning" is suppressed in the teaching of a subject, students do find ways of creating their own sense of what is meaningful. This took place before the language of constructivism had surfaced.

4. See Brown, 1981a for a discussion of Sharon's creation of a totally new algorithm for long division based upon her ostensible misconception. The more general issue of the positive power of confusion is discussed in Brown (1993). The issue is discussed from the point of view of errors and educational significance in chapter 6.

5. Such a claim obscures the agony of a two-thousand-year history—one in which efforts were made to deduce Euclid's fifth postulate from the others. That is, a great deal of problem solving had to take place in order to even imagine that the major difficulty in showing the independence of the postulates was the expectation that they were not independent.

6. See Brown (1991) for a discussion of the pervasive and long-standing nature of problems dealing with the issue of the infinitude of primes. The proof of Euclid that there is an infinite number of prime numbers appears as Proposition 20 of Book IX (Heath, 1956). While I will be offering a numerical conception of the proof, it is worth keeping in mind that Euclid thought of numbers as having "magnitude" and thought of them as line segments—built up from units.

7. See Brown (1991, p. 57) for a quick sketch of how that is established.

8. To journey along these paths is not merely to find a slightly better way of teaching what we had formerly taught. It is also to reconstrue what teaching and learning might be about if the concept of personhood and of self-understanding are taken seriously. For some of these excursions, there may in fact be answers that are acceptable to all. For many the answers will be idiosyncratic and what will be prized is good conversation and the opening up of new terrain for students—terrain that we as educators might only begin to imagine. It is to find a way of viewing curriculum as a vehicle for exploring our uniqueness and not only for coming to understand what has been passed down through the ages. See Sfard (2000) for a discussion of the relationship between an a priori Platonic conception of objects and their creation through discourse.

9. See Maor (1987) for a fine analysis of the relationship of these ideas about infinity to culture in general. Rucker (1983) presents a delightful popularization of these deep ideas.

10. See Brown & Walter (1990, pp. 39–41) for a more elaborate discussion of this fanciful way of defining the problem of the existence of irrationals and of their uncountability.

11. There are many reasons that this is the case. A major one is that each new number system proposed a view of progress that was anti-American (my expression) in the deepest sense. That is, people's notion of progress frequently is that after we have elaborated upon a system, we end up with "more" than what we had before. So people would have been happy if the set of negative numbers had all the properties that the set of natural numbers had but more so. In fact, extending number systems sometimes requires that we no longer enjoy properties in the extended system that we had in the narrower one. The most dramatic example is the extension from real to imaginary numbers—numbers like $\sqrt{-1}$. In the new elaborated system, we no longer enjoy property of order. There is no way of deciding which of two imaginary numbers is larger than the other.

12. This is described under the standard of Mathematical Structure in the National Council of Teachers of Mathematics (1989, pp. 184–186).

13. A succinct way of expressing this fact is to claim that the equivalence relation (a partitioning based upon looking at the remainders of numbers when divided by 3 is uniquely defined with respect to the operation of addition). What we have here is a system of three classes that are called equivalence classes in that they *partition* the set of natural numbers. Thus, the entire system generates the natural numbers (exhaustive), and furthermore, there is no overlap (mutually exclusive) between classes. When we introduce addition as a property we notice something more powerful than this observation. Specifically, from the point of view of addition, it matters not which element we choose from a particular class when we join that element with any other element that also comes from the same equivalence class.

14. This result is more dramatic than it may appear and is discussed in Brown (1966).

15. Of course this analysis raises the question of what we mean by average speed in the first place. What are the circumstances under which it makes sense to get the average of two speeds by adding them up and dividing by 2? If, for example, I travel along a highway for ten hours at 90 miles per hour and then travel for the next five minutes at 10 miles per hour, does it make sense intuitively that my average speed for that four hour and five minute time period would be fifty miles per hour?

16. The concept of different sorts of C_is in mathematical thinking was introduced in Brown & Keren (1972). It has been elaborated considerably in this text. Nelson Goodman (1972) has an excellent discussion of some philosophical problems related to the concept of similarity. Some of his analysis could be viewed as challenging my claim that "same/different" is a powerful mathematical/educational concept. His basic position is that since any two things are always the same (or different) with regard to some specifiable perspective, we cannot use the concept for the purpose of making significant distinctions. Even if correct, however, the claim does not destroy the generative power of "similarity" to help us both create and understand mathematical and nonmathematical worlds.

17. See note 4 chapter 6 for an elaboration of such descriptors.

Chapter 5 An Enhanced View of Connecting Mathematics With the Real World

1. See for example, the co-authored books by Davis & Hersh (1981, 1988), works that have significantly influenced the humanistic mathematics movement.

2. My discussion of the Hersh (1997) essay is adapted from Brown (1997a).

3. Excellent resources that discuss the ubiquitous nature of metaphorical thinking in general are Lakoff & Johnson (1980), Lakoff & Turner (1989), Ortony (1982). Further elaboration of the role they play in mathematics can be found in Brown and Walter (1990), Pimm (1987) and English (1997).

4. The story is adapted from Brown (1974).

5. It might look as if the scheme is overstated in the case of the product of an odd and an even number. This issue as well as a clarification of what precisely is being claimed here is discussed in Brown (1974) and in Brown & Walter (1990, pp. 76-78).

6. It is particularly revealing to compare the metaphors we use in mathematical thinking with those we employ with regard to teaching and learning. In this section I have found the metaphor of striving to be helpful for the purpose of noticing and generating mathematical ideas that derived from "doodling." As I reflect on my mathematical thinking, I have found growth metaphors to be helpful—metaphors that involve striving, birth and identity. I have however, found other metaphors that are a bit more "militaristic" like "zapping." Thus I find it helpful to think of the role of zero under the operation of multiplication as one of imposing itself on any element (zero or otherwise) so as to destroy its identity and to form a collection of like-minded "fellows." See Brown, 1981c. This view is particularly helpful as I try to "hold on to" the more general concept of a group homomorphism. Some of these metaphors are ones that I use in thinking about teaching and learning as well. I find "striving" and "giving birth" to be learning/teaching metaphors that are compatible with my view of mathematics. I find militaristic ones, however, to be at odds with my views of teaching and learning. How I reconcile the use of incompatible metaphors in the two domains poses an interesting problem—the sort of problem that teachers and students might profit from exploring. One interesting observation is that despite the dissimilarity, they are all dynamic metaphors, ones that imply growth of some sort. An excellent resource for coming to as clearer understanding of the kinds of metaphors we use is a classic by Pepper (1942). It would be enlightening to make use of his scheme in an attempt to understand the metaphors we use in Table 1.1, the comparison of progressive and traditional education.

7. I have modified the form of this problem so that it might be a bit more self-referential with regard to the purposes of this book. I received the following watered down (perhaps a tad less sexist?) version of it on February 1997 via the internet from Amy Wardrop. The version I received was the following:

This guy is collecting information for a census, and he goes to a woman's house and rings the doorbell. When the woman answers, he explains that he is collecting information for a census and he needs to know the age of her three daughters. The woman replies, "The product of my 3 daughters' ages equals 72. And the sum of their ages equals the address which you can see on the house." She points to the street number on the house.

The man says, "That is not enough information for me to figure out your daughters' three ages." She replies, "Oh, my oldest daughter has a cat with a wooden leg." The man says, "Thank you. Have a nice day." Based on this information, how old are the woman's three daughters?

Hint: If you are climbing the walls, think of the possibility of twins as an unexplored option. Once you do, come up with as many ways as you can that twins might satisfy the condition that the product of their ages is 72. Notice what sums you end up with for all triples (not triplets!). Then you will see why a comment about her oldest daughter is relevant.

8. I analyzed this problem with a different intention (a focus on some problematic issues related to heuristics of problem solving) in Brown (1985).

9. Some of what follows is an expansion of issues raised in Brown & Lukinsky (1970, 1979).

10. This section is adapted from Brown (1984). A modified version was reproduced in Brown (1986b). Among those who introduced that perspective from a psychological point of view in the 1980s are Gilligan (1982); Belenky, Clinchy, Goldberger, & Tarule (1986). Martin (1981) offers a philosophical analysis. See also Burton (1986b) for a variety of perspectives on mathematics education from a feminist perspective.

11. This question, as well as many others like it, testify to a reconstructed view of what constitutes intelligence. See Gardner (1993, 1999a).

12. After reading some of these theories, the reader may wish to examine the two jokes of Max Black to see what theory(ies) they exhibit. For further discussion of the various philosophical theories and dimensions of humor, see Brown & Brown (1985), Cohen (1999), Morreall (1987), and Paulos (1985). Cohen includes an analysis of humor's moral dimensions.

13. If the Russell paradox seems a bit abstract and technical, consider the following analogous paradox:

There is a town in which the barber shaves all those people and only those people who do not shave themselves. Now, who shaves the barber?

If he shaves himself, then he cannot do so since he is to shave only those people who do not shave themselves. If he does not shave himself, then he shaves himself since he is to shave those people who do not shave themselves.

14. Those who are persuaded that the scheme of mathematical induction is nonproblematic might wonder about why it is necessary to establish P_1 since the second condition appears to be strong enough to do the job. I suggest the following conjecture about the sum of the integers in order to indicate that P_1 is necessary and also to suggest that the entire enterprise is more interesting than appears after one has been "acclimated" to its use:

$$1 + 2 + 3 + 4 + \ldots + n = [n \cdot (n + 1)] / 2 + 17.$$

Now using the same kind of scheme for the second condition of mathematical induction we used to prove the valid equality, see what happens when you try to prove the above. The results will be surprising.

15. Actually, in 1937, Vinogradov improved a bit on Schnirelman's first crack. The results, however, are far from straightforward. That is, he showed that for any "sufficiently large" even number, the sum of at most four primes is necessary. Kramer (1970) discusses the context in which this proof was produced.

Chapter 6 Problems and the Real World: Conclusion and Expansion

1. The discussion in this section is based upon Brown (1968). For further elaboration on the context of the structure movement more generally, see Brown (1979).

2. Actually, it looks as if more is at stake here. It appears on the surface that the axiomatic approach involves legerdemain of sorts. It appears as a "trick" that is totally unmotivated. In fact, it is possible to motivate the use of the distributive property by thinking about what is done by using a pattern inductively, and then transforming that viewpoint into an axiomatic one. See Brown (1978).

3. The school of philosophy of mathematics known as intuitionism challenges the notion of existence that we have been describing. It claims that existence statements are meaningless unless they contain a method for actually constructing the object whose existence is being asserted. It is associated with L. E. J. Brouwer (1881–1967) and Leopold Kronecker (1823–1891). At bottom the school of intuitionism denies the law of the excluded middle with infinite sets. The law states that for any proposition, p, either it or its negation, $\sim p$, must be true.

4. Nozick's (1990) "unmatrixed" list includes the following: "value, meaning, importance, weight, depth, amplitude, intensity, height, vividness, richness, wholeness, beauty, truth, goodness, fulfillment, energy, autonomy, individuality, vitality, creativity, focus, purpose, development, serenity, holiness, perfection, expressiveness, authenticity, freedom, infinitude, enduringness, eternity, wisdom, understanding, life, nobility, play, grandeur, greatness, radiance, integrity, personality, loftiness, idealness, transcendence, growth, novelty, expansiveness, originality, purity, simplicity, preciousness, significance, vastness, profundity, integration, harmony, flourishing, power, and destiny" (p. 182).

5. See Pinar (1998) for a festschrift dedicated to Maxine Greene in honor of her eightieth birthday. Many of the essays in the festschrift borrow from and elaborate upon the concept of "wide-awakeness" and apply it to a variety of disciplines.

6. See Brown (1976) for an elaboration of the confusion between "coming to know" and knowledge. The disinclination to encourage false beliefs is very closely connected with our disinclination to disentangle the two.

7. See, for example Brown, Cooney & Jones (1990); Cooney (1985); Webb & Romberg (1994).

8. See Brown (1997b) for further exploration of the concept. Though not cast explicitly in the language of "thinking like a mathematician," much of critical mathematics education seeks to confront that world-view. For an excellent summary of the movement see Skovsmose & Nielsen (1996).

9. See Stevenson (2000) for a discussion of the relationship between a democratic school environment and its impact on teachers' focus on classroom learning.

10. See Brown (1996a). Following a brief description of the novella, I will be selecting several excerpts from it. They are reprinted by permission of Heinemann, a division of Reed Elsevier Inc., Portsmouth, NH, and the author.

11. See Schiro (1997) for an excellent analysis of how stories can function in the context of mathematical meaning-making. He also summarizes some appealing stories written for elementary school students. Also Borasi, Siegel, Fonzi, & Smith (1998) discuss three different models of reading and point out how the transactional model is most consistent with a constructivist perspective. See also Borasi & Siegel (2000).

12. See, for example, Emihovich (1995); Hopkins (1994); McEwan & Egan (1995).

13. See Paulos (1998) for a discussion of other ways in which the narrative format and mathematics relate to each other.

14. The figure from which it was adapted appeared in Lukinsky (1987). Permission to print the variation is granted by The Yale Law Journal Company and Fred B. Rothman & Company.

15. Well . . . almost all members. Given a somewhat sexist caste system, until recently this study was reserved for males *only* as soon as they reached the age of 13. The movie, *Yentl* (Streisand, 1983) depicts a young woman (acted by Barbra Streisand) who is so taken by Talmudic study that she pretends to be a male in order to qualify to participate.

16. Though not directed to Talmudic study per se, Goldenberg (1993) describes other valuable educational roles that might be incorporated in a Talmudic model.

17. I am grateful to Frederick Reiner and Thomas Giambrone for assisting me in teaching this material.

References

Abbott, Edward A. (1963). *Flatlands: A romance of many dimensions.* New York: Barnes & Noble.

Agre, Gene P. (1982). The concept of problem. *Educational Studies, 13* (2), 121–142.

Anderson, John R., Reder, Lynne M., & Simon, Herbert A. (1996). Situated learning and education. *Educational Researcher, 25*(4), 5–11.

Appelbaum, Peter M. (1995). *Popular culture, educational discourse and mathematics.* Albany: State University of New York Press.

Apple, Michael W. (1992). Do the standards go far enough? Power, policy and practice. *Journal for Research in Mathematics Education, 23*(5), 412–431.

Belenky, M., Clinchy, B. M., Goldberger, N. R., & Tarule, J. N. (1986). *Women's ways of knowing: The development of self, voice, and mind.* New York: Basic Books.

Bishop, Alan, Clements, Ken, Kilpatrick, Jeremy, Laborde, Colette, & Keitel, Christine. (Eds.). (1996). *International handbook of mathematics education.* The Netherlands: Kluwer Academic Press.

Bloor, David. (1991). *Knowledge and social imagery.* Chicago: University of Chicago Press.

Borasi, Raffaella. (1986). *On the educational uses of errors: Beyond diagnosis and remediation.* Unpublished doctoral dissertation, State University of New York at Buffalo.

Borasi, Raffaella. (1987). Exploring mathematics through the analysis of errors. *For the Learning of Mathematics, 7*(3), 1–8.

Borasi, Raffaella, & Brown, Stephen I. (1985). A novel approach to texts. *For the Learning of Mathematics, 5*(1), 21–23.

Borasi, Raffaella, & Siegel, Marjorie. (2000). *Reading counts: Expanding the role of reading in mathematics classrooms.* New York: Teachers College Press.

Borasi, Raffaella, Siegel, Marjorie, Fonzi, Judith, & Smith, Constance. (1998). Using transactional reading strategies to support sense-making and discussion in mathematics classrooms: An exploratory study. *Journal for Research in Mathematics Education, 29*(3), 275–305.

Boyer, C. (1968). *History of mathematics.* Boston: John Wiley and Sons.

Brown, Stephen I. (1966). Implications of an operation uniquely defined with respect to an equivalence relation: The 'freeing' of vectors. *The Mathematics Teacher, 59*(2), 115–123.

Brown, Stephen I. (1968). On bottlenecks in mathematics education. *Teachers College Record, 70* (3), 199–212.

Brown, Stephen I. (1974). Musing on multiplication. *Mathematics Teaching, 61,* 26–30.

Brown, Stephen I. (1976). Discovery and teaching a body of knowledge. *Curriculum Theory Network, 5*(3), 191–218.

Brown, Stephen I. (1978). The product of signed numbers: Dissection of an unmotivated proof. *New York State Mathematics Teacher Journal, 28*(2), 67–70.

Brown, Stephen I. (1979). Some limitations of the structure movement in mathematics education: The meaning of *why*. *Mathematics Gazette of Ontario, 17*(3), 35–40.

Brown, Stephen I. (1981a). Sharon's *Kye*. *Mathematics Teaching, 94,* 11–17.

Brown, Stephen I. (1981b). Problem posing: The problem generation gap. *Math Lab Matrix, 16,* 1–5.

Brown, Stephen I. (1981c). Ye shall be known by your generations. *For the Learning of Mathematics, 1*(3), 27–36.

Brown, Stephen I. (1982). Teaching *whys* and wise teaching. *Teaching Philosophy, 5*(2), 125-133.

Brown, Stephen I. (1984). The logic of problem generation: From morality and solving to de-posing and rebellion. *For the Learning of Mathematics, 4*(1), 9-20.

Brown, Stephen I. (1985). Problem solving and teacher education: The humanism twixt models and *muddles*. In E. Jacobsen (Ed.), *Studies in Mathematics Education*, (Vol. 4, pp. 3-29). Paris: UNESCO.

Brown, Stephen I. (1986a). Liberal education and problem solving: Some curriculum fallacies. In David Nyberg (Ed.). *Proceedings of the Philosophy of Education Society* (pp. 299-311). Normal, IL: Philosophy of Education Society.

Brown, Stephen I. (1986b). The logic of problem generation: From morality and solving to de-posing and rebellion. In Leone Burton (Ed.), *Girls into maths go* (pp. 196-222). Sussex, England: Holt, Rinehart and Winston.

Brown, Stephen I. (1990). The right to be wrong. *For the Learning of Mathematics, 10*(3), 37-38.

Brown, Stephen I. (1991). *Some "prime" comparisons*. Reston, VA: National Council of Teachers of Mathematics.

Brown, Stephen I. (1996a). *Posing mathematically*. Portsmouth, NH: Heinemann Press.

Brown, Stephen I. (1996b). Towards humanistic mathematics education. In Alan Bishop et al. (Eds.), *International handbook in mathematics education* (pp. 1289-1321). The Netherlands: Kluwer Academic Press.

Brown, Stephen I. (1997a). Math lingo vs. plain English: Multiple entendre. *Humanistic Mathematics Network Journal, 15*, 5-10.

Brown, Stephen I. (1997b). Thinking like a mathematician: A problematic perspective. *For the Learning of Mathematics, 17(2)*, 36-38.

Brown, Stephen I., & Brown, Jordan D. (1985). [Review of the book *Mathematics and humor*], *Thinking: The Journal of Philosophy for Children, 6*(1), 52-56.

Brown, Stephen I., Cooney, Thomas J., & Jones, Douglas. (1990). Mathematics teacher education. In W. Robert Houston (Ed.), *Handbook* of research on teacher education (pp. 639–656). New York: Macmillan.

Brown Stephen I., & Keren, Gideon. (1972). Problems in gestalt psychology and traditional logic: A new role for analysis in the doing of mathematics. *Association of Teachers of Mathematics of New England, 5* (1), 5–13.

Brown Stephen I., & Lukinsky, Joseph S. (1970). Morality and the teaching of mathematics. *New York Ethical Culture Society, 1*(2), 2–12.

Brown Stephen I., & Lukinsky, Joseph S. (1979). Integration of religious studies and mathematics in the day school. *Jewish Education,* (Fall), 28–36

Brown, Stephen I., & Walter, Marion, I. (1970). What if not? An elaboration and second illustration. *Mathematics Teaching, 51,* 9–17.

Brown, Stephen I., & Walter, Marion, I. (1977). Problem posing and problem solving: An illustration of their interdependence. *Mathematics Teacher, 70* (1), 4–13.

Brown, Stephen I., & Walter, Marion, I. (1983). *The art of problem posing*. Second edition (1986), Philadelphia: The Franklin Institute Press. Subsequently published (1990) in Hillsdale, NJ: Lawrence Erlbaum.

Brown, Stephen I., & Walter, Marion, I. (1988). Problem posing in mathematics education. *Questioning Exchange, 2*(2), 123–131.

Brown, Stephen I., & Walter, Marion, I. (Ed.). (1993). *Problem posing: Elaborations and applications.* Hillsboro, NJ: Lawrence Erlbaum.

Brownell, William A. (1935). Psychological considerations in the learning and the teaching of arithmetic. *The teaching of arithmetic* (pp. 1–31). New York: Teachers College Press, Columbia University.

Brownell, William A. (1945). When is arithmetic meaningful? *Journal of Educational Research, 38,* 481–498.

Bruner, Jerome. (1986). *Actual minds, possible worlds.* Cambridge, MA: Harvard University Press.

Buerk, Dorothy. (1982). An experience with some able women who avoid mathematics. *For the Learning of Mathematics, 3* (2), 19–24.

Burke, Kenneth. (1968). *Language as symbolic action: Essays on life, literature, and method.* Berkeley: University of California Press.

Burton, Leone. (1986a). *Thinking things through: Problem solving in mathematics.* Oxford: Basil Blackwell.

Burton, Leone. (Ed.). (1986b). *Girls into maths go.* Sussex, England: Holt, Rinehart and Winston.

Butts, Thomas. (1980). Posing problems properly. In Stephen Krulik & Robert E. Reys (Eds.), *Problem solving in school mathematics: 1980 Yearbook* (pp. 23–32). Reston, VA: National Council of Teacher of Mathematics.

Cohen, Felix. (1929). What is a question? *The Monist, 39,* 350–364.

Cohen, Ted. (1999). *Jokes: Philosophical thoughts on joking matters.* Chicago: University of Chicago Press.

Comiti, Claude, & Ball, Deborah. (1996). Preparing teachers to teach mathematics; A comparative perspective. In Alan Bishop et al. (Eds.). *International handbook of mathematics education* (pp. 1123–1153). The Netherlands: Kluwer Academic Press.

Cooney, Thomas J. (1985). A beginning teacher's view of problem solving. *Journal for Research in Mathematics Education, 16*(5), 324–336.

Davis, Philip J., & Hersh, Reuben. (1981). *The mathematical experience.* Boston: Birkhäuser.

Davis, Philip J., & Hersh, Reuben. (1988). *Descartes' dream.* London: Penguin Books.

Dawson, John W. (1999). Gödel and the limits of logic. *Scientific American, 280*(6), 76–81.

Dennett, Daniel C. (1990). The interpretation of texts, people and other artifacts. *Philosophy and Phenomenological Research, 50,* supplement, 177–194.

Dewey, John. (1902). *The child and the curriculum.* Chicago: University of Chicago Press.

Dewey, John. (1910). *How we think.* Boston: D. C. Heath.

Dewey, John. (1920). *Reconstruction in philosophy.* Boston: Beacon Press.

Dewey, John. (1975). *Moral principles in education.* Carbondale: Southern Illinois University Press.

Dewey, John. (1988). Democracy for the teacher. In Stephen I. Brown and Mary Finn (Eds.), *Readings from progressive education: A movement and its professional journal* (pp. 199–201). Lanham, MD: University Press of America.

Dewey, John. (1988). Progressive education and the science of education. In Stephen I. Brown and Mary Finn (Eds.), *Readings from progressive education: A movement and its professional journal* (pp. 160–167). Lanham, MD: University Press of America.

Dillon, James T. (1986). Student questions and individual learning. *Educational Theory, 36* (4), 333–341.

Dossey, John. (1996). *Modeling with functions.* Portsmouth, NH: Heinemann Press.

Dowker, Ann. (1992). Computational estimation strategies of professional mathematicians. *Journal for Research in Mathematics Education, 23*(1), 45–54.

Egan, Kieran. (1979). *Educational development.* New York: Oxford University Press.

Egan, Kieran. (1988). *Teaching as story telling.* London: Routledge.

Emihovich, Catherine. (1995). Distancing passion: Narratives in social science. *Qualitative Studies in Education, 8*(1), 37–48.

English, Lyn. (Ed.). (1997). *Mathematical reasoning: Analogies, metaphors and images.* Mahway, NJ: Lawrence Erlbaum Associates, Inc.

Ernest, Paul. (1991). *The philosophy of mathematics education.* London: The Falmer Press.

Ernest, Paul. (Ed.). (1994a). *Constructing mathematical knowledge: Epistemology and mathematics education.* London: The Falmer Press.

Ernest, Paul. (Ed.). (1994b). *Mathematics, education and philosophy: an international perspective.* London: The Falmer Press.

Ernest, Paul. (Ed.). (1995). *Philosophy of Mathematics Education Newsletter. 8*, Exeter, England: Philosophy of Mathematics Education Network (University of Exeter) England.

Ernest, Paul. (1997). *Social Constructivism as a philosophy of mathematics.* Albany: State University of New York Press.

Fajtlowicz, Siemion. (1988). On conjectures of graffiti. *Discrete Mathematics, 72,* 113–118.

Fawcett, Harold P. (1938). *The nature of proof: A description and evaluation of certain procedures used in a senior high school to develop an understanding of the nature of proof. Thirteenth yearbook of the national council of teachers of mathematics.* New York: Teachers College, Columbia University Press.

Fielker, David. (1990). Observation lessons. *For the Learning of Mathematics, 10*(1), 16–22.

Freire, Paulo. (1970). *Pedagogy of the oppressed.* New York: Seabury Press.

Freud, Sigmund. (1960). *Jokes and their relation to the unconscious.* New York: Norton. (originally published in 1905).

Gardner, Howard. (1993). *Multiple intelligences: The theory in practice.* New York: Basic Books.

Gardner, Howard. (1999a). *Intelligence reframed: Multiple Intelligences for the 21st century.* New York: Basic Books.

Gardner, Howard. (1999b). *The disciplined mind: What all students should understand.* New York: Simon & Schuster.

Gardner, Martin. (1975). *The paradox box.* San Francisco: W. H. Freeman and Company.

Gardner, Martin. (1998, September 24). The new new Math. *The New York Review of Books,* pp. 9–12.

Gay, John, & Cole, Michael. (1967). *The new mathematics and an old culture: A study of learning among the Kpelle of Liberia.* New York: Holt, Rinehart and Winston.

Gilligan, Carol. (1982). *In a different voice.* Cambridge, MA: Harvard University Press.

Gleick, James. (1987). *Chaos: Making a new science.* Fairfield, PA: Penguin Books.

Goldenberg, E. Paul. (1993). On building curriculum materials that foster problem posing. In Stephen I. Brown & Marion I. Walter (Eds.),

Problem posing: Reflections and applications (pp. 31–38). Hillsdale, NJ: Lawrence Erlbaum.

Goodman, Nelson. (1972). *Problems and projects.* Indianapolis: Bobbs-Merrill.

Green, Thomas F. (1971). *The activities of teaching.* New York: McGraw-Hill.

Greene, Maxine. (1978). *Landscapes of learning.* New York: Teachers College Press.

Greenhut, Robert. (Producer), Allen, Woody. (Director). (1985). *The Purple Rose of Cairo* [Film]. Santa Monica: Orion Pictures Corporation.

Grube, G. M. A. (Translator). (1976). *Plato's meno.* Indianapolis: Hackett Publishing Co.

Hallie, Philip. (1979). *Lest innocent blood be shed: The story of the Village of Le Chambon and how goodness happened there.* New York: HarperPerennial.

Hallie, Philip. (1997). *Tales of good and evil, help and harm.* New York: Harper Collins Publishers.

Hattiangadi, J. N. (1978). The structure of problems. *Philosophy of the Social Sciences, 8*, 345–365; *9*, 49–71.

Haug, Mikel. (1998). Up the creek with a paddle. *The Mathematics Teacher, 91*(6), 456–460.

Heath, Thomas, L. (1956). *The thirteen books of Euclid's elements.* New York: Dover Publications.

Henderson, David. (1996). I learn mathematics from my students—Multiculturalism in Action. *For the Learning of Mathematics, 16*(2), 46–52.

Hersh, Reuben (1997). Math lingo vs. plain language. *American Mathematical Monthly, 104* (1), 48–51.

Hirst, Paul H. (1970). Liberal education and the nature of knowledge. In Jane Martin (Ed.), *Readings in the philosophy of education: A study of curriculum* (pp. 157–178). Boston: Allyn and Bacon.

Hirst, Paul H. (1974). *Knowledge and the curriculum: A collection of philosophical papers.* London: Routledge and Kegan Paul.

Hobbes, Thomas. (1914). *Leviathan*. London: Dent. (originally published in 1651).

Hofstadter, Douglas. (1979). *Gödel, Escher, Bach: An eternal golden braid*. New York: Vintage Books.

Hofstadter, Douglas. (1985). Metamathematical themas: Variations on a theme as the crux of creativity. New York: Basic Books.

Hopkins, Richard L. (1994). *Narrative schooling*. New York: Teachers College Press.

Horgan, John. (1993). The death of proof. *Scientific American. 269*(4), 92–110.

Jarvie, I. C. (1979). Laudan's problematic progress and the social sciences. *Philosophy of Social Science, 9*, 484–97.

Keyser, Cassius J. (1916). Humanization of teaching mathematics. In C. J. Keyser (Ed.). *The human worth of rigorous thinking* (pp. 75-89). New York: Columbia University Press.

Kilpatrick, Jeremy. (1987). Problem formulating: Where do good problems come from? In Alan Schoenfeld (Ed.), *Cognitive science and mathematics education*. (pp. 123–147). Hillsdale, NJ: Lawrence Erlbaum.

Kimball, Bruce. (1986). *Orators and philosophers: A history of the idea of a liberal education*. New York: Teachers College Press.

Kincheloe, Joe, Steinberg, Shirley, & Tippins, Deborah. (1992). *The stigma of genius: Einstein and beyond modern education*. Durango, CO: Hollowbrook Publishing.

Kitcher, Philip. (1988). *The nature of mathematical knowledge*. Oxford: Oxford University Press.

Kline, Morris. (1966a). Intellectuals and the schools: A case history. *Harvard Educational Review, 36*(4), 505–511.

Kline, Morris. (1966b). A proposal for the high school mathematics curriculum: What does it mean? *The Mathematics Teacher, 59*(4), 322–334.

Kohlberg, Lawrence. (1976). Moral stages and motivation: The cognitive development approach. In T. Likona (Ed.). *Moral development and behavior: Theory, research and social issues* (pp. 31–53). New York: Holt, Rinehart & Winston.

Kolata, Gina (1989, June 18). A program that makes conjectures. *New York Times Current Events Section,* 47.

Knuth, Donald E. (1974). *The rock: Surreal numbers.* Reading, MA: Addison Wesley.

Kramer, Edna E. (1970). *The nature and growth of modern mathematics.* New York: Hawthorne Books.

Krutetskii, Vadim A. (1976). *The psychology of mathematical abilities in school children* . In Jeremy Kilpatrick and Izaak Wirszup (series Eds.). *Soviet studies in the psychology of learning and teaching mathematics.* Chicago: University of Chicago Press.

Kuhn, Thomas S. (1962). *The structure of scientific revolutions.* Chicago: University of Chicago Press.

La Noue, George R. (1962). Religious schools and secular subjects. *Harvard Educational Review, 32*(3), 255–291.

Lakatos, Imre. (1976). *Proofs and refutations.* Cambridge, England: Cambridge University Press.

Lakoff, George, & Johnson, Mark. (1980). *Metaphors we live by.* Chicago: University of Chicago Press.

Lakoff, George, & Turner, Mark (1989). *More than cool reason: A field guide to poetic metaphor.* Chicago: University of Chicago Press.

Laudan, Larry. (1977). *Progress and its problems.* Berkeley: University of California Press.

Laudan, Larry. (1984). *Science and values.* Berkeley: University of California Press.

Lave, Jean. (1988). *Cognition in practice: Mind, mathematics and culture in everyday life.* New York: Cambridge University Press.

Leacock, Stephen. (1956). Mathematics for golfers. In J. R. Newman (Ed.), *The world of mathematics* (pp. 2456–2459). New York: Simon and Schuster.

Lehman, Hugh. (1977). On understanding mathematics. *Educational Theory, 27*(2), 111–119.

Lehrer, Tom, McLees, David, Hansen, Barry, & Inglott, Bill (Producers). (2000). *The remains of Tom Lehrer.* [Compact disks]. Los Angeles: Warner Bros. Records & Rhino Entertainment Company.

Lipman, Matthew. (1974). *Harry Stottlemeier's Discovery.* Upper Montclair, NJ: Institute for the Advancement of Philosophy for Children.

Lipman, Matthew. (1988). *Philosophy goes to school.* Philadelphia: Temple University Press.

Lipman, Matthew. (1991). *Thinking in education.* Cambridge, England: Cambridge University Press.

Lipman, Matthew, Sharp, Ann, & Oscanyon, Frederick. (1977). *Philosophy in the classroom.* Upper Montclair, NJ: Institute for the Advancement of Philosophy for Children.

Lipman, Matthew, Johnson, Thomas, & Gazzard, Anthony. (1989). *Philosophy for children: Where we are now.* Supplement no. 2. Upper Montclair, NJ: Institute for the Advancement of Philosophy for Children.

Lukinsky, Joseph. (1987). Law in education: A reminiscence with some footnotes to Robert Cover's *Nomos* and *Narrative.* The Yale Law Journal, *96*(8), 1836–1859.

Lyons, Nona, P. (1983). Two perspectives on self, relationships, morality. *Harvard Educational Review, 53*(2), 125–145.

MacLane, Saunders. (1998). Letter to the editor. *Focus, 6*(2), 6.

Mandelbrot, Benoit. (1999). A multifractal walk down Wall Street. *Scientific American, 280*(2), 70–74.

Maor, Eli. (1987). *To infinity and beyond: A cultural history of the infinite.* Boston: Birkhäuser.

Martin, Jane. (1970). *Explaining, understanding and teaching.* New York: McGraw-Hill.

Martin, Jane.(1981). Needed: A new paradigm for liberal education. In Jonas F. Soltis (Ed.). *National society for the study of education* (pp. 37–59). Chicago: University of Chicago Press.

Martin, Jane. (1988). Science in a different style. *American Philosophical Quarterly, 25*(2), 129–140.

Mason, John. (1994). Researching from the inside in mathematics education: Locating an I-You relationship. In *Educational Proceedings of the Eighteenth Annual Conference of the International Group for the Psychology of Mathematics Education* (pp. 176–194). Lisbon, Portugal: University of Lisbon.

McEwan, Hunter, & Egan, Kieran. (Eds.). (1995). *Narrative in teaching, learning and research.* New York: Teachers College Press.

McPeck, John E. (1981). *Critical thinking and education.* New York: St. Martin's Press.

McPeck, John E. (1990). *Teaching critical thinking: Dialogue and dialectic.* New York: Routledge.

McPeck, John E. (1992). Thoughts on subject specificity. In Steven Norris (Ed.), *The generalizability of critical thinking: Multiple perspectives on an educational ideal* (pp. 198–205). New York: Teachers College Press.

Mills, William H. (1947). A prime-representing function. *American Mathematical Society Bulletin,* 53, 604.

Mitroff, Ian, & Kilmann, Ralph. (1978). *Methodological approaches to the social sciences.* San Francisco: Jossey-Bass.

Moïse, Edwin E. (1963). *Elementary geometry from an advanced standpoint.* Reading, MA: Addison-Wesley.

Moïse, Edwin E. (1965). Activity and motivation in mathematics. *American Mathematical Monthly,* 72, 407–412.

Morreall, John. (Ed.). (1987). *The philosophy of laughter and humor.* Albany: State University of New York Press.

Montagu, Ashley. (1981). *Growing Young.* New York: McGraw Hill Book Company.

Nagel, James, & Newman, James. (1958). *Gödel's proof.* New York: New York University Press.

National Council of Teachers of Mathematics. (1989). *Curriculum and evaluation standards for school mathematics.* Reston, VA: Author.

National Council of Teachers of Mathematics. (1991). *Professional standards for teaching mathematics.* Reston, VA: Author.

National Council of Teachers of Mathematics. (1995). *Assessment standards for school mathematics.* Reston, VA: Author.

National Council of Teachers of Mathematics.(2000). *Principles and standards for school mathematics.* Reston, VA: Author.

National Research Council. (1989). *Everybody counts: A report to the nation on the future of mathematics education.* Washington, DC: National Academy Press. Author.

Newman, James. (1956). Srinivasa Ramanujan. In James Newman (Ed.), *The world of mathematics.* (pp. 368-376). New York: Simon and Schuster.

Nickles, Thomas. (1981). What is a problem that we may solve it? *Synthèse, 47,* 85-115.

Nickson, Marilyn. (2000). *Teaching and learning mathematics: A teacher's guide to recent research.* London: Cassell.

Nickson, Marilyn, & Lerman, Stephen (Eds.). (1992). *The social context of mathematics education: Theory and practice.* London: South Bank Press.

Noddings, Nel. (1994). Does everybody count? Reflections on reform in school mathematics, *Journal of Mathematical Behavior, 13,* 89-104.

Nozick, Robert. (1990). *The examined life: Philosophical meditations.* New York: Simon & Schuster.

Nussbaum, Martha. (1990). *Love's knowledge: Essays on philosophy and literature.* New York: Oxford University Press.

Ortony, Andrew. (Ed.). (1982). *Metaphor and thought.* Cambridge, England: Cambridge University Press.

Papy, Frederique. (1975). *The little dreamer.* St. Louis: Central Midwestern Regional Laboratory.

Paulos, John Allen. (1982). *Mathematics and humor.* Chicago: University of Chicago Press.

Paulos, John Allen. (1985). *I think, therefore I laugh: An alternative approach to philosophy.* New York: Columbia University Press

Paulos, John Allen. (1998). *Once upon a number.* New York: Basic Books.

Pepper, Stephen C. (1942). *World Hypotheses: A Study of Evidence.* Berkeley: University of Caifornia Press.

Perry, William G., Jr. (1970). *Forms of intellectual and ethical development in the college years.* New York: Holt, Rinehart and Winston.

Petrie, Hugh G. (1981). *The dilemma of enquiry and learning.* Chicago: University of Chicago Press.

Phenix, Philip. (1964). *Realms of Meaning.* New York: McGraw Hill.

Pimm, David. (1987). *Speaking mathematically: Communication in the mathematics classroom.* New York: Routledge and Kegan, Paul.

Pinar, William F. (Ed.). (1998). *Passionate mind of Maxine Greene.* Bristol, PA: Falmer Press, Taylor & Francis Inc.

Poincaré, Henri. (1961). Mathematical creation in Brewster Ghiselin (Ed.). *The creative process* (pp. 33–44). New York: Menton Books. (original work published in 1913)

Polya, Georg. (1954). *Mathematics and plausible reasoning (Vols. 1–2).* Princeton, NJ: Princeton University Press.

Polya, Georg. (1957). *How to solve it.* New York: Anchor Books.

Polya, Georg. (1962). *Mathematical discovery: On understanding, learning and teaching problem solving* Vols. 1–2. New York: John Wiley & Sons.

Popper, Karl. (1959). *The logic of scientific discovery.* New York: Basic Books.

Raimi, Ralph, & Braden, Lawrence. (1998). *State mathematics standards: An appraisal of math standards in 46 states, the District of Columbia and Japan.* Washington, DC: Thomas B. Fordham Foundation.

Rapaport, William J. (1982). Unsolvable problems and philosophical progress. *American Philosophical Quarterly, 19*(4), 289–298.

Restivo, Sal, Van Bendegem, Jean Paul, & Fischer, Roland. (Eds.). (1993). *Math worlds: Philosophical, and social studies of mathematics and mathematics education.* Albany: State University of New York Press.

Rogers, Robert. (1964). Mathematical and philosophical analyses. *Philosophy of Science Journal, 31*, 255–264.

Rosen, Jonathan. (1998). The Talmud and the internet. *The Key Reporter, 63*(4), 8–11.

Rosen, Jonathan. (2000). *The Talmud and the internet: A journey between worlds.* New York: Farrar, Strauss & Giroux.

Rucker, Rudy. (1983). *Infinity and the mind: The science and philosophy of the infinite.* Toronto: Bantom Books.

Russell, Bertrand. (1908). Mathematical logic as based on the theory of sets. *American Journal of Mathematics, 30*, 222–262.

Sarason, Seymour. (1971). *The Culture of school and the problem of change.* Boston: Allyn and Bacon.

Scheffler, Israel. (Ed.). (1958). *Philosophy and education.* Boston: Allyn and Bacon.

Scheffler, Israel. (1960). *The language of education.* Springfield, IL: Charles C. Thomas.

Scheffler, Israel. (1965). *Conditions of knowledge: An introduction to epistemology and education.* Glenview, IL: Scott, Foresman.

Scheffler, Israel. (1967). *Science and subjectivity.* Indianapolis: Bobbs-Merrill.

Scheffler, Israel. (1973). Reflections on educational relevance. In Israel Scheffler (Ed.), *Reason and teaching* (pp. 126–135). Indianapolis: Bobbs-Merrill Co. Inc.

Scheffler, Israel. (1985). *Of human potential: An essay in the philosophy of education.* Boston: Routledge & Kegan Paul.

Scheffler, Israel. (1991). *In praise of the cognitive emotions.* New York: Routledge.

Schiro, Michael. (1997). *Integrating children's literature and mathematics in the classroom: Children as meaning makers, problem solvers, and literary critics.* New York: Teachers College Press.

Schlick, Moritz. (1935). Unanswerable questions? *The Philosopher, 13.* Reprinted in Michael Meyer (Ed.). (1988). *Questions and questioning.* Berlin: Walter de Gruyter.

Schoenfeld, Alan. (1980). Heuristic behavior variables in instruction. In Gerald A. Goldin & C. Edwin McClintock. (Eds.), *Task variables in mathematical problem solving*, (pp. 431–461). Columbus, OH. ERIC Clearinghouse for Science, Mathematics and Environmental Education.

School Mathematics Study Group. (1961). *First course in algebra*. New Haven: Yale University Press. Author

Sfard, Anna. (2000). Steering (dis)course between metaphors and rigor: Using focal analysis to investigate an emergence of mathematical objects. *Journal for Research in Mathematics Education, 31*(3), 296–327.

Sierpinska, Anna. (1990). Some remarks on understanding in mathematics. *For the Learning of Mathematics, 10*(3), 24–41.

Silver, Edward A. (1994). On mathematical problem posing. *For the Learning of Mathematics, 14*(1), 19–28.

Silver, Edward A., Mamona-Downs, Joanna, Leung, Shukkwan S., & Kenney, Patricia Ann. (1996). Posing mathematical problems. *Journal for Research in Mathematics Education,* 27(3), 293–309.

Sintonen, Matti. (1985). Separating problems from their backgrounds: A question-theoretic proposal. *Communication and Cognition, 18* (1–2), 25–49.

Skovsmose, Ole, & Nielsen, Lene. (1996). Critical mathematics education. In Alan Bishop et al. (Eds.), *International Handbook in Mathematics Education* (pp. 1257–1288). The Netherlands: Kluwer Academic Press.

Steen, Lynn. (Ed.). (1990a). *On the shoulders of giants*. Washington, DC: National Academy Press.

Steen, Lynn. (1990b). Mathematics for all Americans. In Thomas J. Cooney (Ed.), *Teaching and learning mathematics in the 1990s: 1990 yearbook,*. Reston, VA: National Council of Teachers of Mathematics.

Stein, Sherman. (1996). The triex: Explore, extract, explain. *Humanistic Mathematics Network. 14,* 6–8.

Stevenson, Robert B. (2001). Shared decision-making and core school values: A case study of organizational learning. *International Journal of Educational Management*, special issue, forthcoming.

Stor, Marilyn, & Briggs, William L., (1998). Dice and disease in the classroom. *The Mathematics Teacher, 91*(6), 464-468.

Streisand, Barbra, & Lemorande, Rusty. (Producers), Streisand, Barbra (Director). (1983). *Yentl* [Film]. Santa Monica: Metro-Goldwyn-Mayer, Inc./United Artists Entertainment.

Toulmin, Stephen. (1977). *Human understanding.* Princeton: Princeton University Press.

Tymoczko, Thomas. (1985). *New directions in the Philosophy of mathematics.* Boston: Birkhäuser.

Tymoczko, Thomas. (1986). Making room for mathematicians in the philosophy of mathematics. *Mathematical Intelligencer, 8,* 44-50.

Tymoczko, Thomas. (1993), Humanistic and utilitarian aspects of mathematics. In A. White (Ed.), *Essays in humanistic mathematics.* Washington DC: Mathematical Association of America.

University of Illinois Committee on School Mathematics. (1961). *Arithmetic of real numbers* (Vol. 1). Urbana: University of Illinois Press. Author.

Van Bendegem, Jean Paul. (1993). Foundations of mathematics or mathematical practice: Is one forced to choose? In S. Restivo, J. P. Van Bendegem, R. Fisher (Eds.), *Math worlds: Philosophical and social studies of mathematics and mathematics education* (pp. 21-38). Albany: State University of New York Press.

Verducci, Susan. (2000). A conceptual history of empathy and a question it raises for moral education. *Educational Theory, 50*(1), 63-80.

Walter, Marion I., & Brown, Stephen I. (1969). What if not? *Mathematics Teaching, 46,* 38-45.

Webb, Norman L., & Romberg, Thomas A. (Eds.). (1994). *Reforming mathematics education in America's cities.* New York: Teachers College Press.

Wertime, Richard. (1979). Students, problems and "courage spans." In J. Lockhead & J. Clements (Eds.), *Cognitive process instruction* (pp. 27-36). Philadelphia: The Franklin Institute Press.

Weyl, Hermann. (1951). A half-century of mathematics. *American Mathematical Monthly, 58*(8), 523-553.

White, Alvin. (Ed.). (1993). *Essays in humanistic mathematics*. Washington, DC: The Mathematical Association of America.

Zeigarnik, Bliuma V. (1965). *The pathology of thinking*. New York: Consultants Bureau.

Index

Abbott, Edward A., 214
abstraction
    concrete commonalty and, 184–188
    concrete incompatibility and, 188–189
Agre, Gene P., 70–72, 75, 80, 97
Allen, Woody, 17, 176
Anderson, John R., 239
Appelbaum, Peter M., 11
Apple, Michael W., 11
applications, 8–11, 139, 141, 142, 153, 181–182, 193, 215
    mathematical models, 8–10, 16, 139–143
    model variations, 142–144
    shared non-mathematical experience, 144–146, 166–180, 195–196
Aristotle, 53, 78, 168, 212
Art of Problem Posing, The (Brown/Walter), 238
Assessment Standards for School Mathematics (NCTM), 236
axiomatics, 3, 19, 23, 185, 187, 196

Back to Methuselah (Shaw), 52
Ball, Deborah, 236
Belenky, Mary, 244
Bergson, Henri, 168
Bishop, Alan, 236
Black, Max, 166, 167
Bloor, David, 23
Bolyai, Jonas, 49, 222

Borasi, Raffaella, 201, 206, 246
Boyer, C., 222
Braden, Lawrence, 11
Briggs, William L., 9
Brouwer, L.E.S., 245
Brown, Jordan D., 83, 86, 95, 244
Brown, Stephen I., 44, 64, 94, 179, 200, 206, 236, 237, 238, 239, 240, 241, 242, 243, 244, 245, 246
Bruner, Jerome, 216
Brownell, William A., 240
Buerk, Dorothy, 18–19
Burke, Kenneth, 50–51
Burton, Leone, 235
Butts, Thomas, 235

caring, 164–166
chaos theory, 171–173
Child and the Curriculum, The (Dewey), 237
Clements Ken, 236
Clinchy, Blythe M., 244
clock arithmetic, 117–121
Cohen, Felix, 73, 74, 75
Cohen, Ted, 244
Cole, Michael, 204, 214
Comiti, Claude, 236
computer, 8, 79, 174
    irony of, 24–27
    computer science, 37, 171
constructivism, 10, 21, 26, 41, 85, 98, 194, 241, 246

Cooney, Thomas J., 246
creativity, 17, 113-114, 202, 236
   counter-factual and, 52-53
   understanding and, 56
   What-If-Not and, 55-56
Curriculum and Evaluation Standards for School Mathematics (NCTM), 6, 8, 45, 236

Davis, Philip J., 25, 243
Dawson, John W., 237
definition
   imposing of, 62-64
   meaning of, 240
   understanding purpose of, 62-64
discovery
   "new math" and, 4
   verification and, 49
distributive property
   geometry and algebra, 219
   hidden abstraction and, 184-188
   isomorphism, 123-125
   "new math" and, 4
   Talmud and, 227-233
Dennett, Daniel C., 79-80
Dewey, John, 7, 10, 26, 27, 53, 57, 71, 87, 204, 237, 240
Dillon, James T., 73
Dossey, John, 140
Dowker, Ann, 25

educational fallacy, 18, 31, 70, 76-80, 83, 97, 194
Egan, Kieran, 18, 213, 214, 246
Einstein, Albert, 129, 192, 219
Elements, The (Euclid), 222, 238
Emihovich, Catherine, 246
English, Lyn, 243
Erlanger program, 94
Ernest, Paul, 24, 239
errors, 6, 42, 128
error-making and generative power, 86, 95, 103, 198-201, 218
Escher, M.C., 171
Euclid, 23, 40, 46
   Euclidean geometry, 23, 40, 238, 241
   non-Euclidean geometry, 47, 94, 198, 222
   parallel postulate and, 46-49, 94, 222, 238
   proof of infinitude of primes, 99-102, 189, 241
Euclid Alone Has Looked on Beauty (Millay), 16
Euclides Ab Omne Naevo Vindicatus (Saccheri), 47
Everybody Counts: A Report to the Nation on the Future of Mathematics Education (Author), 12, 236
existence of mathematical objects, 17, 108, 110, 147, 148

Fajtlowicz, Siemion, 25
Fawcett, Harold P., 40-41
Federman, Raymond, 176
feminist epistemology, 49, 197, 238, 244
Fielker, David, 199
Fischer, Roland, 24
Flatland (Abbott), 214
Fonzi, Judith, 246
fractals, 173-175
Freud, Sigmund, 55, 113, 153, 168
function, 7, 9, 73, 139, 189, 194

game of jeopardy, 12-15, 69, 194
Gardner, Howard, 9
Gardner, Martin, 9
Gay, John, 204-214
Gauss, Carl, 34, 35, 36, 155, 162
Gazzard, Anthony, 211
genius
   idiot savant, 39
   mathematical talent, 4, 39-40, 201-202
   problem solving and, 36-39, 93, 100
   reconceiving a field and, 94
Gilligan, Carol, 163, 164, 244
Gleick, James, 173, 174
Gödel, Kurt, 23, 93, 173, 237

Gödel, Escher, Bach: An Eternal Golden Braid (Hofstadter), 171, 213
Goldbach, C., 180, 181, 197
Goldberger, N.R., 244
Goldenberg, E. Paul, 246
Goodman, Nelson, 242
Green, Thomas F., 105, 11, 112
Greene, Maxine, 191, 246
Greenhut, Robert, 17, 176
Grube, G.M.A., 213, 240

Hallie, Philip, 212
Hardy, David, 38, 39
Harry Stottlemeier's Discovery (Lipman), 211
Hattiangadi, J.N., 75
Haug, Mikel, 9
Heath, Sir Thomas, L., 46, 222, 238, 241
Henderson, David, 112
Hersh, Reuben, 25, 146, 147, 243
heuristics, 6, 11, 69, 77, 244
Hilbert, David, 23, 93
Hirst, Paul H., 236
Hobbes, Thomas, 168
Hofstadter, Douglas, 52, 53, 171, 213, 237
Hopkins, Richard L., 246
Horgan, John, 25
How We Think (Dewey), 71
humanistic
 ambiguity of the concept of, 29-30
 hidden mind and, 20
 isolation from experience, 21
 mathematics education, 18-30, 215
 mathematics network, 22, 27-28
 narrow view of logic, 20
 novel and. *See* novel as pedagogical tool
 perspective, 18-29, 221, 236
 progressive education and, 29-30
 radical perspective, 29, 194, 201, 202
 Talmud and. *See* Talmud
Humanistic Mathematics Newsletter, 28

humor
 abstraction and, 169-170
 "Black" humor, 166-167
 ingenuity and, 168-169
 math is funny, 178-181
 nature of, 167-168
 self-referentiality and, 170-178

intentionality, 83-88, 96
intuition, 27, 41, 106-110, 129, 161, 174, 219, 236
isomorphism
 clock arithmetic, 117-121
 distributive property, 123-125
 equilateral triangles, 121-123
 equivalence classes, 117-121

Jarvie, I.C., 240
Johnson, Mark, 243
Johnson, Thomas, 211
Jokes and Their Relation to the Unconscious (Freud), 168
Jones, Douglas, 246

Keitel, Christine, 236
Kenney, Patricia Ann, 235
Keren, Gideon, 242
Keyser, Cassius J., 153
Kilmann, Ralph, 183, 200, 215
Kilpatrick, Jeremy, 236
Kimball, Bruce, 236
Kincheloe, Joe, 39
Kitcher, Philip, 24
Klein, Felix, 94
Kline, Morris, 185-188, 236
Knuth, Donald E., 213
Koestler, Arthur, 168
Kohlberg, Lawrence, 162-164
Kolata, Gina, 25
Kramer, Edna E., 245
Kronecker, Leopold, 245
Krutetskii, Vadim A., 39, 201-202
Kuhn, Thomas S., 49

La Noue, George R., 153
Laborde, Colette, 236
Lakatos, Imre, 24, 40, 82, 93, 213

Lakoff, George, 243
language and metaphor, 146–148
Laudan, Larry, 238
Lave, Jean, 239
Leacock, Stephen, 213
Lehman, Hugh, 57, 239
Lehrer, Tom, 131, 170
Lerman, Stephen, 24
Leung, J., 235
Leviathan, The (Hobbes), 168
Lipman, Matthew, 211, 214
Little Dreamer, The (Papy), 213
Lobachevsky, N. I., 49
Lukinsky, Joseph, 216, 218, 244
Lyons, Nona, P., 165

MacLane, Saunders, 11
Mamona-Downs, J., 235
Mandelbrot, Benoit, 173–175
Maor, Eli, 241
Martin, Jane, 64, 237, 238, 244
Marx, Groucho, 170
Marx, Sharon, 170, 241
Mason, John, 239
Mathematics and Humor (Paulos), 168, 170
Mathematics for Golfers (Leacock), 214
Mathematics Teacher, The (Journal), 9
McEwan, Hunter, 246
McPeck, John E., 11
meaning
 construction of, 10, 240
 devoid of, 169–170
 language ambiguity and, 146–148, 153
 problem and, 83, 86–88
 problem solving and, 86
 scientific, 54, 55
 students and, 87, 237
 subject matter and , 97
 two senses of, 86, 88, 229
Meno, The (Plato), 74, 213
metaphor
 irony in mathematics and, 152
 mathematics and, 146–152, 243
 non-mathematical contexts, 181, 195, 243
 teaching/learning and, 243
Mills, William H., 178, 188
Mitroff, Ian, 183, 205, 215
Moïse, Edwin E., 42, 47
Molière, 168
Moore, Robert Lee, 42–43
morality
 feminist perspective on, 162–166
 principled perspective on, 153–162
Morreall, John, 244
Montagu, Ashley, 18

Nagel, James, 237
negative thinking, 50–55
"new math," 3–5, 19, 20, 28, 40 43, 116, 170, 184, 196, 204
New Mathematics and an Old Culture, The: A Study of Learning Among the Kpelle of Liberia (Gay and Cole), 204
Newman, James, 39, 237
Nickles, Thomas, 74–76, 96, 97
Nickson, Marilyn, 8, 24
Nielsen, Lene, 246
Noddings, Nel, 12
noticing, 38, 52, 61, 66, 91, 92, 95–97, 112, 125, 239, 243
novel as pedagogical tool, 176, 183, 204–215, 238, 246
Nozick, Robert, 17, 189, 190, 245
number systems
 imaginary, 44, 147, 213. 242
 irrational, 84, 109, 110, 147, 148, 191, 213, 242
 language of, 146–148
 negative (*numeri ficti*), 59, 147, 184–187, 242
 rational, 108, 109
 real, 179, 242
Nussbaum, Martha, 212

Ortony, Andrew, 243
Oscanyon, Frederick, 211

Papy, Frederique, 213
Paulos, John Allen, 168–169, 246
Pepper, Stephen C., 243
Perry, William G. Jr., 76
Petrie, Hugh G., 240
Phenix, Philip, 237
philosophy of mathematics, 22–24, 29
Philosophy of Mathematics Education (Newsletter), 24, 239
Piaget, Jean, 10
Pimm, David, 243
Pinar, William F., 246
Plato, 53, 212, 213, 240, 241
Poincaré, Henri, 56, 57, 239
Polya, Georg, 6
Popper, Karl, 54, 55, 57
Posing Mathematically (Brown), 208, 235
prime number
   different domains and, 58–62
   Goldbach's conjecture, 180–181, 197
   infinitude of primes, 99–102, 189, 241
   understanding and, 57–62
Principles and Standards for School Mathematics (NCTM), 236
problem
   concept of a, 70–73, 194
   constraints and, 74–75, 95–97
   demand and, 74–75
   educational fallacy and. *See* educational fallacy
   the *given* and, 20, 70
   history of a, 95–97
   humanistic conception of, 80, 164, 194, 200
   meaning and, 83, 85–87
   purpose and, 31, 66, 78, 79, 82, 83–86
   self-understanding and, 111, 193, 241
   situation and, 88–93
problem posing
   implicit in problem solving, 43, 45
   philosophical issues, 50–55
   problem's history, 46

   problem solving and, 36, 43, 44, 45, 69
   purpose and, 194
   significance of, 45–49
   What-If-Not and. *See* What-If-Not
Problem Posing: Elaborations and Applications (Brown/Walter), 238
problem solving
   centrality in mathematics, 36–40
   centrality in mathematics education, 5–7, 12–15
   college scene, 42–43
   context and, 11, 88, 93, 98
   See educational fallacy
   purpose and, 21, 51, 62–64, 194
   redefining a field and, 94–95
   scope and, 83–88, 96, 240
   secondary school example, 40–41
Professional Standards for Teaching Mathematics (NCTM), 236
progress in mathematics
   problem solving and, 93–94
   redefining a field and, 93–94
progressive education, 22, 26–27
   Progressive Education Association, 26
   relation to humanistic perspective, 29
proof, 3, 8, 22–24, 106–108, 177–178
Proofs and Refutations (Lakatos), 213
pseudo-history, 95, 196–198, 200, 213, 218
Purple Rose of Cairo, The (Allen), 17, 176
Pythagorean
   theorem, 222
   triples, 43–45

question
   answer and, 72, 76
   meaningfulness and, 172
   presumptions and, 74
   significance and, 73–74
   validity and, 74

Questioning Exchange (journal), 73

Raimi, Ralph, 11
Ramanujan, Srinivasa,, 38–39
real world, 9–10, 17, 153–181, 243
    183–182, 245
Rapaport, William J., 76, 80
Reder, Lynne M., 239
Restivo, Sal, 24
Rock, The: Surreal Numbers
    (D. E. Knuth), 213
Rogers, Robert, 110
Romberg, Thomas A., 246
Rosen, Jonathan, 215, 217
Rucker, Rudy, 224
Russell, Bertrand, 172, 173, 192, 244

Saccheri, Gerolomo, 47
Sarason, Semour, 3, 11, 204
Scheffler, Israel, 23, 78, 104, 238,
    239, 240
Schiro, Michael, 246
Schlick, Moritz, 75, 240
Schnirelman, Lev, 181, 245
Schoenfeld, Alan, 11
School Mathematics Study Group,
    184–185
self-referentiality, 170–177, 192, 208,
    237
self-understanding, 102, 111, 166,
    237, 241
Sfard, Anna, 241
Sharp, Ann, 211
Shaw, George Bernard, 52
Siegel, Marjorie, 246
Sierpinska, Anna, 239
Silver, Edward A., 235
similarity, 113–135, 196
    deep similarity, 134
    formal similarity, 134
    same/different and human nature,
        114–116
    surface similarity, 133–134
    Talmud and, 227–233
    three problems and, 89–91, 126–
        127
    "with regard to," 133–134

Simon, Herbert A., 239
Sintonen, Matti, 75
situation
    neutralizing problems, 191, 194
    problem and, 88–93
    problem and solution as *given*,
        103–104
Skovsmose, Ole, 246
Smiles on Washington Square
    (Federman), 176
Smith, Constance, 246
social context of education, 10, 11, 27,
    202, 237
solution
    concept of problem and, 70–76
    as given, 97–104
Standards, The (NCTM), 5, 29, 36,
    45, 69, 170, 192, 193, 196
Steen, Lynn, 5, 236
Stein, Sherman, 235
Steinberg, Shirley, 39
Stevenson, Robert B., 246
Stor, Marilyn, 9
Streisand, Barbra, 246
structure in mathematics, 17, 23, 38–
    39, 86, 116, 169, 186

Talmud, 31, 183, 215
    distributive property and, 227–233
    explanation and interpretation, 216
    Gemora, 215, 216, 220–222
    internet, 215
    Mishna, 215, 216, 220–222
    secular, 215–222
    typical page, 218
Tarule, J.N., 244
thinking like a mathematician, 201–
    202, 246
Tippins, Deborah, 39
Toulmin, Stephen, 94
Turner, Mark, 243
Tymoczko, Thomas, 24

understanding
    creativity and. *See* creativity and
        What-If-Not
    definition and, 62–64

meaning and, 57, 86, 239
prime number and, 57–62
University of Illinois Committee on School Mathematics, The (Author), 4

Van Bendegem, Jean Paul, 24
Verducci, Susan, 160
Vinogradov, I.M., 245
Vygotsky, Lev, 10

Walter, Marion I., 44, 64, 235, 238, 241, 243
Webb, Norman L., 246
Wertime, Richard, 80
Weyl, Hermann, 94
What-If-Not
  educational matters, 55–62
  foreground/background and, 221
  intention and, 66
  meaning and, 56, 85–86
  metaphor, 151
  philosophical issues, 50–55
  problem posing and, 31, 35, 50–55, 69, 165, 180, 193
  problem solving and, 149, 152
  stages, 64–67
  understanding and, 57–62, 239
  wonder and, 112
White, Alvin, 27, 237
wonder, 21, 104–106, 110–113, 161, 169, 192, 212
  fundamental theorem of integral calculus, 106–107
  infinite orchard problem, 109–110

Yentl (Streisand), 246

Zeigarnik, Bluima V., 234